Computational Geometric Algebra for Physicists With Python

Jamie Flux

Contents

1 Scalars and Vectors: The Building Blocks of GA   **14**
    Scalar Quantities and Their Properties   14
    Vector Quantities: Definition and Fundamental Operations   14
    1   Computational Implementation of Vectors . .   15
    Operations Involving Scalars and Vectors   15
    Python Code Snippet   16

2 Multivectors: Combining Grades Beyond Vectors   **21**
    Definition and Structure of Multivectors   21
    Algebraic Formation and Operations on Multivectors   21
    Computational Representation of Multivectors . . .   22
    Python Code Snippet   23

3 Grades and Blades: Understanding the Structure   **28**
    Grades in Geometric Algebra   28
    Blades as Fundamental Components   28
    Hierarchical Structure and Computational Representation   29
    Python Code Snippet   30

4 The Geometric Product: Definition and Computation   **35**
    Algebraic Definition of the Geometric Product . . .   35
    Properties and Structural Implications   36
    Computational Strategies for the Geometric Product   36
    Python Code Snippet   38

5 Inner Product in GA: Concepts and Formulations   **42**
    Algebraic Foundations of the Inner Product   42

  Comparative Analysis: Inner Product and Traditional Dot Product 43
  Physical Significance of the Inner Product 43
  Computational Realization of the Inner Product Operation 44
  Python Code Snippet 45

6 Outer Product: The Wedge Product Essentials   **48**
  Geometric Interpretation of the Wedge Product . . . 48
  Algebraic Structure and Construction of Higher-Grade Elements 49
  Computational Implementation of the Wedge Product 50
  Python Code Snippet 51

7 Clifford Algebras: Foundations for Computational Physics   **56**
  Mathematical Foundations 56
  Algebraic Structure and Product Operations 56
  Computational Representation and Algorithmic Implementation 57
  Relevance in Modeling Physical Phenomena 59
  Python Code Snippet 59

8 Basis and Representation: Constructing the Algebra   **63**
  Basis Choices in Geometric Algebra 63
  Coordinate Representations and Computational Implications 64
  Explicit Construction of Multivectors for Computational Tasks 64
  Python Code Snippet 65

9 Reversion Operation: Properties and Computation   **70**
  Mathematical Definition of Reversion 70
  Properties of the Reversion Operation 71
  Computational Implementation of Reversion 71
  Python Code Snippet 73

10 Grade Involution: Signatures and Effects   **77**
  Mathematical Formulation of Grade Involution . . . 77
  Algebraic Impact of Grade Involution 78
  Computational Implementation of Grade Involution 78
  Python Code Snippet 79

11 Clifford Conjugation: Techniques and Applications 84
    Mathematical Foundations of Clifford Conjugation . 84
    Algebraic Simplification and Symmetry Analysis . . 85
    Computational Implementation of Clifford Conjugation 85
    Python Code Snippet 86

12 Norms in Geometric Algebra: Measuring Magnitudes 91
    Mathematical Framework of Norms in Geometric Algebra 91
    Physical Significance of Norms 92
    Computational Implementation of Norms in Geometric Algebra 92
    Python Code Snippet 93

13 Inversion of Multivectors: Computational Methods 98
    Mathematical Framework for Multivector Inversion . 98
    Algorithmic Considerations and Inversion Criteria . 99
    Computational Implementation of Multivector Inversion 99
    Python Code Snippet 100

14 Exponential Functions in GA: Theory and Computation 105
    The Theoretical Foundation 105
    Decomposition and Closed-Form Expressions 106
    Computational Considerations 106
    Python Implementation of a Multivector Exponential Function 107
    Python Code Snippet 107

15 Logarithms in GA: An Analytical Perspective 112
    Foundations of Multivector Logarithms 112
    Series Expansion and Convergence Considerations . 113
    Branch Structure and Multivector Decomposition . . 113
    Computational Strategies for Evaluating $\log(M)$ . . 114
    Python Implementation of the Logarithmic Function 114
    Python Code Snippet 115

16 Rotors: Generators of Rotations in GA — 121
Mathematical Framework of Rotors in Geometric Algebra 121
Construction and Properties of Rotors 122
Implementation in Computational Frameworks . . . 122
Python Implementation of Rotor Application 123
Python Code Snippet 123

17 Reflections: Representation through Geometric Algebra — 127
Mathematical Framework of Reflections in GA . . . 127
Derivation of Reflection Formulas 128
Physical Interpretations of Reflection Operations . . 128
Python Implementation: Reflection Function 129
Python Code Snippet 129

18 Spinors: Their Role in Geometric Physics — 132
Mathematical Foundations of Spinors 132
Spinors and Rotational Symmetries 132
Computational Approach to Spinor Rotations 133
Spinor Algebra in the Geometric Framework 134
Python Code Snippet 134

19 Rotations in GA: From Theory to Computation — 139
Fundamental Principles of Rotations in Geometric Algebra 139
Computational Implementation of Rotational Transformations 140
Python Code Snippet 141

20 Lorentz Transformations: GA Approach to Relativity — 146
Foundations of Lorentz Transformations in Geometric Algebra 146
Exponential Map for Boosts and Hyperbolic Dynamics 147
Computational Implementation of Lorentz Boosts in GA 147
Python Code Snippet 149

21 Complex Structures in GA: Embedding and Computation — 152
Embedding of Complex Numbers in Geometric Algebra 152

  Computational Representation and Efficiency 153
  Applications in Computational Physics and Signal
  Processing 154
  Python Code Snippet 154

22 Matrix Representations: Linking GA and Linear Algebra   158
  Matrix Representations of GA Elements 158
  Matrix Operators and GA Multiplications 159
  Algorithmic Implementation of GA Matrix Mappings 159
  Python Code Snippet 161

23 Algorithms for the Geometric Product: Implementation Strategies   164
  Mathematical and Computational Foundations . . . 164
  Bitwise Representations and Table-Driven Approaches 165
  1 Bitwise Computation of the Sign Factor . . . 165
  Data Structures and Algorithmic Optimizations in
  Geometric Algebra 166
  1 Efficient Implementation of the Geometric
    Product for Multivectors 166
  Python Code Snippet 167

24 Symbolic Manipulation in GA: Tools and Techniques 171
  Foundations of Symbolic Computation in Geometric
  Algebra 171
  Software Tools and Libraries Supporting Symbolic GA 172
  Algorithmic Strategies for Symbolic Simplification . 172
  1 Function for Symbolic Simplification 173
  Python Code Snippet 173

25 Basis Transformations: Changing Perspectives in GA   178
  Mathematical Foundations of Basis Transformations 178
  Representation of Multivectors Under Basis Change 179
  Computational Strategies for Basis Transformations 179
  Implementation Example: A Basis Transformation
  Function 180
  Python Code Snippet 181

26 Duality in Geometric Algebra: Theoretical Foundations 185
Mathematical Basis of Duality 185
Physical Interpretations and Significance 186
Computational Implementation of Duality in Geometric Algebra 186
Python Code Snippet 187

27 Projectors: Extracting Subspace Components 193
Mathematical Formulation of Projection Operators . 193
Algorithmic Considerations in Projection Operations 194
Properties and Applications of Projection Operators 195
Python Code Snippet 195

28 Subspace Representation: Blades and Their Geometry 202
Blades as Algebraic Representatives of Subspaces . . 202
Geometric Interpretation of Blades 202
Algorithmic Handling of Blades in Computational Systems 203
Python Code Snippet 204

29 Intersection Operations: Computing Geometric Intersections 208
Algebraic Framework for Subspace Intersections . . . 208
Intersection via Duality: The Meet Operation 209
Computational Considerations in Intersection Operations 210
Python Code Snippet 210

30 Dirac Algebra in GA: Bridging Quantum Theory 213
Algebraic Foundations 213
Representation of Gamma Matrices and Spinors . . . 214
Implications for Quantum Physics 214
Computational Implementation: Python Function for the Dirac Operator in GA 215
Python Code Snippet 215

31 Reformulating Maxwell's Equations with GA: A New Perspective 219
Geometric Algebra Formulation of Electromagnetism 219
Unified Expression of Maxwell's Equations in GA . . 220
Computational Implementation in Python 220

  Python Code Snippet 221

32 Equations of Motion: GA in Dynamical Systems   224
  Formulating Dynamics in Geometric Algebra 224
  The Geometric Derivative and Differential Equations 225
  Application to Rigid Body Dynamics 225
  Computational Implementation for GA-based Dynamics 226
  Python Code Snippet 227

33 GA in Quantum Mechanics: A Unified Language   231
  The Algebraic Structure of Quantum States 231
  Observables and Operators in Geometric Algebra . . 232
  Quantum Dynamics and the Geometric Product . . 232
  Computational Considerations in GA Quantum Mechanics 233
  Python Code Snippet 233

34 Wave Functions as Multivectors: Representation and Computation   239
  Multivector Formulation of Quantum Wave Functions 239
  Transformation and Invariance Properties 240
  Computational Efficiency and Benefits 240
  Representative Implementation Function 241
  Python Code Snippet 241

35 Operator Theory in GA: Linear and Nonlinear Aspects   246
  Operator Formalism in Geometric Algebra 246
  Linear Operator Methods 247
  Nonlinear Operator Techniques 248
  Operator Commutation and Algebraic Structures . . 249
  Python Code Snippet 249

36 Angular Momentum in GA: Computational Representations   254
  Mathematical Foundations 254
  Computational Representation of Angular Momentum 254
  Algorithmic Considerations and Analysis 256
  Python Code Snippet 256

37 Pauli Algebra: Linking Spin and Geometry — 260
Algebraic Structure and Mapping in Geometric Algebra 260
Spinor Representation and Rotor Construction ... 261
Computational Considerations in Pauli-Algebraic Operations 262
Python Code Snippet 263

38 Commutators in GA: Algebraic Structures Explored — 267
Mathematical Definition and Algebraic Properties . 267
Physical Relevance in Symmetry and Dynamics ... 268
Computational Implementation of Commutators .. 268
Python Code Snippet 270

39 Anti-Commutators: Properties and Physical Interpretations — 274
Mathematical Formulation and Algebraic Characteristics 274
Implications in Symmetry and Transformation Analysis 275
Computational Implementation of Anti-Commutators in Geometric Algebra 275
Python Code Snippet 276

40 Differential Operators in GA: A Computational Framework — 281
Mathematical Foundations of Differential Calculus in Geometric Algebra 281
The Vector Derivative in Geometric Algebra 282
Algorithmic Strategies for Computing Derivatives of Multivector Fields 282
Computational Considerations and Efficiency 283
Python Code Snippet 284

41 Multivector Derivatives: Algorithms and Applications — 288
Mathematical Formulation of Multivector Derivatives 288
Algorithmic Implementation of Multivector Differentiation 289
Applications in Computational Modeling 290
Python Code Snippet 291

42 Differential Forms in GA: Integration and Differentiation — 295
Mathematical Foundations of Differential Forms in GA 295
Integration in the Framework of Geometric Algebra 296
Differentiation through the Exterior Derivative . . . 296
Algorithmic Implementation: Computation of the Exterior Derivative 297
Python Code Snippet 298

43 Lie Groups in GA: Connecting Algebra with Geometry — 303
Foundations of Lie Groups and Geometric Algebra . 303
Representing Lie Groups in Geometric Algebra . . . 304
Code Implementation and Computational Aspects . 304
Algebraic Structure and Geometric Transformations 305
Analysis of Commutators and the Exponential Map in GA 306
Python Code Snippet 306

44 Lie Algebras: Structure and Symmetry in GA — 312
Algebraic Foundations of Lie Algebras in GA 312
Properties and Structural Aspects of the Lie Bracket 313
Computational Techniques for Lie Algebra Operations 313
Representation of Physical Symmetries in GA 314
Python Code Snippet 315

45 Symmetry Transformations: Representing Physics with GA — 319
Theoretical Foundations of Symmetry Transformations 319
Computational Methods for Generating Rotors . . . 320
Applications in Physical Systems 321
Python Code Snippet 322

46 Conservation Laws in GA: Deriving Invariants — 326
Algebraic Foundations and Invariant Structures . . . 326
Derivation of Invariants from Geometric Structures . 327
Computational Approaches to GA Invariants 327
Python Code Snippet 329

47 Noether's Theorem in GA: Mathematical Insights — 332
Algebraic Reformulation of Noether's Theorem in Geometric Algebra 332
Symmetry Transformations and GA Invariants . . . 333
Computational Implementation of GA Noether Invariants 333
Python Code Snippet 334

48 Gauge Theory Foundations in GA: A Unified Treatment — 337
Mathematical Formulation of GA Gauge Fields . . . 337
Representation of Gauge Connections Using Geometric Algebra 338
Computational Implementation of Gauge Transformations in GA 338
Python Code Snippet 339

49 Spin Representations: Computation with GA — 344
Mathematical Formalism of Spin Representations . . 344
Computational Techniques for Spin Transformations 345
Spinor Dynamics and Computational Operations . . 346
Python Code Snippet 346

50 Phase Space in GA: Representation and Simulation — 350
Multivector Structure of Phase Space 350
Computational Strategies for Phase Space Simulation 351
1 Python Implementation of a Phase Space Update Function 351
Python Code Snippet 352

51 Fourier Transforms in GA: Computational Techniques — 357
Fundamentals of Fourier Analysis in Geometric Algebra 357
Mathematical Formulation and Exponential Representation 358
Computational Implementation of the Fourier Transform in GA 358
Python Code Snippet 359

52 Function Spaces and GA: Bridging Analysis and Geometry — 363
Integration of Function Spaces in Geometric Algebra 363
Operators and Algebraic Structures in Function Spaces 364

Analytic Expansion and Basis Representations . . . 365
Computational Implementation of GA Function Spaces 365
Python Code Snippet 366

53 Integration Techniques: GA Methods for Computation 371
Numerical Integration Methods in Geometric Algebra 371
Symbolic Integration Techniques in GA Function Spaces 372
Algorithmic Frameworks for GA Integration 373
Python Code Snippet 374

54 Geometric Calculus: Differentiation within GA 379
Foundations of Differentiation in Geometric Algebra 379
Symbolic Differentiation Techniques in Geometric Algebra 380
Numerical Differentiation Approaches in Geometric Algebra 381
Python Code Snippet 382

55 The Laplacian in GA: Computation and Applications 385
Definition of the Laplacian in Geometric Algebra . . 385
Computational Implementation of the Laplacian . . 386
Applications in Physical Systems 387
Python Code Snippet 388

56 Electromagnetic Fields in GA: Modeling and Analysis 392
Mathematical Formalism of Electromagnetic Field Representation 392
Computational Representation of Electromagnetic Field Bivectors 393
Analysis of Electromagnetic Field Dynamics in Geometric Algebra 394
Python Code Snippet 395

57 Relativistic Mechanics with GA: Computation in Spacetime 399
Relativistic Spacetime Algebra 399
Lorentz Transformations and Boost Rotors 399
Computational Implementation of Lorentz Boosts . . 400
Spacetime Dynamics through GA Computational Operations 401

        Python Code Snippet 401

58 Computational Implementations: Designing GA Algorithms   404
        Foundations of Multivector Representation 404
        Algorithmic Implementation of the Geometric Product 405
        Runtime Efficiency and Precision Management . . . 407
        Python Code Snippet 407

59 Optimizing the Geometric Product: Efficiency in Computation   413
        Precomputation Strategies 413
        Data Structure Optimization 414
        Algorithmic Enhancements 415
        Python Code Snippet 416

60 Numerical Methods in GA: Tackling Complex Problems   420
        Foundations of Numerical Computation in Geometric Algebra 420
        Discretization and Representation Schemes 421
        Iterative Solvers and Convergence Techniques 421
        Adaptive Integration Methods for Multivector Fields 422
        Error Propagation and Stability Analysis 422
        Python Code Snippet 423

61 Symbolic Algorithms: Automating GA Computations   428
        Foundations of Symbolic Computation in Geometric Algebra 428
        Automated Simplification Techniques in Geometric Algebra 429
        Algorithmic Implementation and Efficiency Considerations 430
        Python Code Snippet 430

62 Visualization of Multivectors: Techniques for Physicists   435
        Foundations of Multivector Visualization 435
        Dimensional Reduction and Projection Techniques . 436
        Computational Techniques for Rendering Multivectors 436
        Integration with Visualization Frameworks 437
        Python Code Snippet 438

63 Multidimensional Rotations: Exploring Higher Dimensions 442
Mathematical Foundations of Higher-Dimensional Rotations 442
Representation of Higher-Dimensional Rotors in Geometric Algebra 443
Algorithmic Construction of Multidimensional Rotations 443
Python Code Snippet 444

64 Tensor and Multivector Interplay: Algebraic Connections 448
Mathematical Foundations 448
Algebraic Mappings and Interplay 449
Computational Conversion and Implementation . . . 449
Interconversion of Representational Formalisms . . . 450
Python Code Snippet 451

65 Data Structures in GA: Efficient Computational Representations 454
Design Principles and Memory Layout 454
Mapping Algebraic Structures to Data Types 455
Implementation of Custom GA Data Structures . . . 455
Efficiency and Optimization Considerations 456
Python Code Snippet 457

66 Simulation Techniques with GA: Tools for Physical Modeling 462
Mathematical Framework for GA-Based Simulations 462
Discretization and Time Integration in GA Simulations 463
Algorithmic Implementation Example 463
Python Code Snippet 464

Chapter 1

Scalars and Vectors: The Building Blocks of GA

Scalar Quantities and Their Properties

Within the framework of geometric algebra, scalar quantities are defined as elements of grade 0. Scalars are typically identified with elements of the real numbers, denoted by $\mathbb{R}$. Their algebraic operations—addition, subtraction, and multiplication—adhere to the familiar rules of commutativity and associativity. For any two scalars α and β, the relation
$$\alpha\beta = \beta\alpha$$
holds true. In physical systems, scalars often represent invariant quantities such as mass or temperature, which, regardless of the coordinate system, maintain their numerical value during transformations.

Vector Quantities: Definition and Fundamental Operations

Vectors are the elements of grade 1 within geometric algebra. A vector is characterized by both magnitude and direction and is commonly represented in an n-dimensional space as

$$\mathbf{v} = v^1\mathbf{e}_1 + v^2\mathbf{e}_2 + \cdots + v^n\mathbf{e}_n,$$

where $\{\mathbf{e}_i\}$ constitutes an orthonormal basis. Physically, vectors are employed to model directional quantities such as velocity, force, and displacement. Their geometric interpretation, coupled with their algebraic properties, makes them indispensable in modeling the dynamics of physical systems.

1 Computational Implementation of Vectors

In computational contexts, the representation of vectors necessitates careful design to balance efficiency with clarity of mathematical operations. A straightforward approach involves encapsulating the components of a vector within a sequential data structure. The following Python function exemplifies a minimal implementation that constructs a vector from a variable number of components:

```
def create_vector(*components):
    """
    Constructs a vector from provided components.

    Each argument represents the coordinate along a basis direction.
    The function returns a list, thereby preserving the sequential
    order of components. This representation directly supports
        vector
    addition and scalar multiplication operations.
    """
    return list(components)
```

This function utilizes a variable-length argument list to accommodate vectors of arbitrary dimensionality. The representation as a list is chosen for its simplicity and for ease of performing element-wise operations.

Operations Involving Scalars and Vectors

The interaction between scalars and vectors forms the cornerstone of geometric algebra. Scalar multiplication is defined such that when a scalar α multiplies a vector $\mathbf{v}$, the result is given by

$$\alpha\mathbf{v} = (\alpha v^1)\mathbf{e}_1 + (\alpha v^2)\mathbf{e}_2 + \cdots + (\alpha v^n)\mathbf{e}_n.$$

This operation scales the vector uniformly while preserving its direction.

Vector addition is computed by summing corresponding components in a chosen basis. For two vectors **v** and **w** expressed in the same orthonormal basis,

$$\mathbf{v} + \mathbf{w} = (v^1 + w^1)\mathbf{e}_1 + (v^2 + w^2)\mathbf{e}_2 + \cdots + (v^n + w^n)\mathbf{e}_n.$$

This operation is both commutative and associative, ensuring that the vector space structure remains robust under coordinate transformations.

Furthermore, the geometric product exhibits compatibility with scalar multiplication. In situations where one factor is a scalar, the product satisfies the relation

$$\alpha\mathbf{v} = \mathbf{v}\alpha.$$

This property upholds the consistency of algebraic manipulation across different elements of geometric algebra and reinforces the perspective that scalars act as a natural scaling mechanism within the algebraic structure.

Python Code Snippet

```
#!/usr/bin/env python3
# -*- coding: utf-8 -*-
"""
This module demonstrates the basic operations on scalars and vectors
within the context of geometric algebra. It includes:
  - Creation of vectors.
  - Vector addition.
  - Scalar multiplication.
These operations correspond to the equations:
  v + w = (v1 + w1, v2 + w2, ..., vn + wn)
  * v = (*v1, *v2, ..., *vn)
and the property that for any scalar  and vector v,
  * v = v * .
"""

def create_vector(*components):
    """
    Constructs a vector from the provided components.

    Args:
        *components: The coordinate values along each basis
            direction.
```

```
        Returns:
            A list representing the vector.
        """
        return list(components)

    def add_vectors(v, w):
        """
        Adds two vectors component-wise.

        Args:
            v: A vector represented as a list of numbers.
            w: A vector represented as a list of numbers.

        Returns:
            A new vector which is the element-wise sum of v and w.

        Raises:
            ValueError: If the vectors are not of the same dimension.
        """
        if len(v) != len(w):
            raise ValueError("Vectors must be of the same dimension for
            ↪  addition.")
        return [a + b for a, b in zip(v, w)]

    def scalar_multiply(scalar, v):
        """
        Multiplies a vector v by a scalar.

        Args:
            scalar: A number representing the scalar.
            v: A vector (list of numbers).

        Returns:
            A new vector where each component of v is multiplied by the
            ↪  scalar.
        """
        return [scalar * x for x in v]

    def display_vector(name, v):
        """
        Utility function to display a vector with a given name.

        Args:
            name: String representing the name of the vector.
            v: The vector to display.
        """
        print(f"{name} = {v}")

# Advanced representation using a Vector class with operator
↪  overloading.
class Vector:
        """
```

```python
    A simple vector class for representing grade 1 elements in
    ↪ geometric algebra.
    """
    def __init__(self, *components):
        """
        Initializes the vector with the provided components.

        Args:
            *components: Coordinate values along each basis
                ↪ direction.
        """
        self.components = list(components)

    def __add__(self, other):
        """
        Returns the vector sum of self and another Vector.

        Args:
            other: A Vector instance to be added.

        Returns:
            A new Vector representing the sum.

        Raises:
            TypeError: If other is not a Vector.
            ValueError: If the vectors have different dimensions.
        """
        if not isinstance(other, Vector):
            raise TypeError("Can only add another Vector.")
        if len(self.components) != len(other.components):
            raise ValueError("Vectors must be of same
                ↪ dimensionality.")
        return Vector(*(a + b for a, b in zip(self.components,
            ↪ other.components)))

    def __rmul__(self, scalar):
        """
        Implements scalar multiplication when the scalar comes
        ↪ first.

        Args:
            scalar: A numeric value.

        Returns:
            A new Vector where each component is multiplied by the
                ↪ scalar.
        """
        return Vector(*(scalar * x for x in self.components))

    def __mul__(self, scalar):
        """
        Implements scalar multiplication when the vector is on the
        ↪ left.
```

```python
        Args:
            scalar: A numeric value.

        Returns:
            A new Vector scaled by the scalar.

        Raises:
            TypeError: If the multiplier is not a scalar.
        """
        if isinstance(scalar, (int, float)):
            return Vector(*(scalar * x for x in self.components))
        raise TypeError("Multiplication only supports scalars.")

    def __str__(self):
        """
        Returns a string representation of the vector.
        """
        return f"Vector({', '.join(map(str, self.components))})"

    def __repr__(self):
        return self.__str__()

if __name__ == "__main__":
    # --------------------------------------------------
    # Demonstration using list-based vector operations.
    # --------------------------------------------------
    # Create vectors using the functional approach.
    v = create_vector(2, 3, 4)
    w = create_vector(1, 0, -1)

    display_vector("v", v)
    display_vector("w", w)

    # Vector addition: v + w = (2+1, 3+0, 4+(-1))
    v_plus_w = add_vectors(v, w)
    display_vector("v + w", v_plus_w)

    # Scalar multiplication: 3 * v = (3*2, 3*3, 3*4)
    scaled_v = scalar_multiply(3, v)
    display_vector("3 * v", scaled_v)

    # --------------------------------------------------
    # Demonstration using the Vector class with operator
    #   overloading.
    # --------------------------------------------------
    vec1 = Vector(2, 3, 4)
    vec2 = Vector(1, 0, -1)

    print("\nUsing the Vector class:")
    print("vec1 =", vec1)
    print("vec2 =", vec2)
```

```python
# Vector addition using overloaded '+' operator.
sum_vector = vec1 + vec2
print("vec1 + vec2 =", sum_vector)

# Scalar multiplication using overloaded '*' operator.
scaled_vector = 3 * vec1   # Equivalent to vec1 * 3
print("3 * vec1 =", scaled_vector)

# Verify commutativity of scalar multiplication.
scaled_vector_commutative = vec1 * 3
print("vec1 * 3 =", scaled_vector_commutative)
```

Chapter 2

Multivectors: Combining Grades Beyond Vectors

Definition and Structure of Multivectors

A multivector in geometric algebra is an element composed of a direct sum of homogeneous components, each associated with a distinct grade. In mathematical form, an arbitrary multivector M can be expressed as

$$M = \langle M \rangle_0 + \langle M \rangle_1 + \langle M \rangle_2 + \cdots,$$

where $\langle M \rangle_r$ denotes the grade-r component. Unlike traditional vectors that inhabit a linear space of grade 1, multivectors encapsulate elements from grade 0 (scalars), grade 1 (vectors), grade 2 (bivectors), and so forth. Through this construction, multivectors serve as a natural extension of conventional vector spaces, capturing subspace elements such as areas, volumes, and higher-dimensional analogues in a unified algebraic framework.

Algebraic Formation and Operations on Multivectors

The synthesis of multivectors relies on the linear combination of elements across different grades. Each homogeneous component is associated with a specific basis generated by the outer (wedge) product. The sum of two multivectors A and B is defined grade-wise:

$$A + B = \sum_r \left(\langle A \rangle_r + \langle B \rangle_r \right).$$

Operations such as the geometric product extend the familiar dot product and cross product from vector algebra. The geometric product of multivectors is associative and distributive over addition, and it generally produces a mixture of terms of different grades. This algebraic structure not only preserves but also enriches the geometric interpretation of traditional operations, permitting rigorous descriptions of rotations, reflections, and projections within a single formalism.

Computational Representation of Multivectors

In computational implementations, a robust representation of multivectors must accommodate their intrinsic heterogeneous structure. A common approach employs a data structure in which each grade is associated with its corresponding components. The use of a dictionary is advantageous because it naturally maps grades (such as 0, 1, 2, etc.) to their numeric or vectorial representations. This organization facilitates grade-wise arithmetic operations, including addition and scalar multiplication, by applying these operations to the components stored under each grade key.

A demonstrative function in Python constructs a multivector as a dictionary. The scalar component is assigned to key 0, the vector component to key 1, and a bivector component to key 2. The function is presented as follows:

```
def create_multivector(scalar, vector=None, bivector=None):
    """
    Constructs a multivector representation using a dictionary.
    The scalar part is assigned to key 0; if provided, the vector
        part is
    stored with key 1, and the bivector part with key 2.

    Args:
        scalar: The grade-0 component, representing the scalar.
        vector: The grade-1 component, typically a list representing
            a vector.
```

```
        bivector: The grade-2 component, represented as a list
        ↪  capturing the
            bivector elements in context.

    Returns:
        A dictionary representing the multivector with grades as
        ↪  keys.
    """
    mv = {0: scalar}
    if vector is not None:
        mv[1] = vector
    if bivector is not None:
        mv[2] = bivector
    return mv
```

The function exemplifies the encapsulation of distinct grade elements within a unified data structure. By segregating the different grades, algebraic operations can be performed in a targeted and efficient manner, preserving the hierarchical and combinatorial features of geometric algebra.

Python Code Snippet

```
def create_multivector(scalar, vector=None, bivector=None):
    """
    Constructs a multivector representation using a dictionary.
    The scalar part is stored under key 0, the vector part under key
    ↪  1,
    and the bivector part under key 2.

    Args:
        scalar: The grade-0 component (a number).
        vector: The grade-1 component (typically a list representing
        ↪  a vector).
        bivector: The grade-2 component (represented as a list for
        ↪  bivector elements).

    Returns:
        A dictionary representing the multivector.
    """
    mv = {0: scalar}
    if vector is not None:
        mv[1] = vector
    if bivector is not None:
        mv[2] = bivector
    return mv

def add_components(x, y):
```

```python
    """
    Adds two components, which can be numbers or lists.

    Args:
        x: First component.
        y: Second component.

    Returns:
        The sum of the two components.
    """
    if isinstance(x, list) and isinstance(y, list):
        return [xi + yi for xi, yi in zip(x, y)]
    return x + y


def add_multivectors(mv1, mv2):
    """
    Adds two multivectors grade-wise.

    Args:
        mv1: First multivector (dictionary).
        mv2: Second multivector (dictionary).

    Returns:
        A new multivector representing the grade-wise sum.
    """
    result = {}
    all_grades = set(mv1.keys()).union(mv2.keys())
    for grade in all_grades:
        comp1 = mv1.get(grade, 0)
        comp2 = mv2.get(grade, 0)
        result[grade] = add_components(comp1, comp2)
    return result


def scalar_mult_component(s, comp):
    """
    Multiplies a scalar with a component, which can be a number or
    ↪ list.

    Args:
        s: Scalar multiplier.
        comp: The component (number or list).

    Returns:
        The product of s and comp.
    """
    if isinstance(comp, list):
        return [s * x for x in comp]
    return s * comp


def scalar_multiply_multivector(s, mv):
```

```python
    """
    Multiplies every component of a multivector by a scalar.

    Args:
        s: Scalar multiplier.
        mv: Multivector (dictionary).

    Returns:
        A new multivector with scaled components.
    """
    result = {}
    for grade, comp in mv.items():
        result[grade] = scalar_mult_component(s, comp)
    return result


def dot_product(v1, v2):
    """
    Computes the dot (inner) product between two vectors.

    Args:
        v1: First vector (list).
        v2: Second vector (list).

    Returns:
        The scalar dot product.
    """
    return sum(x * y for x, y in zip(v1, v2))


def wedge_product(v1, v2):
    """
    Computes the wedge (outer) product between two 3D vectors.
    For 3-dimensional vectors, the bivector is represented as a list
    corresponding to the basis elements [e12, e13, e23].

    Args:
        v1: First vector (list with 3 elements).
        v2: Second vector (list with 3 elements).

    Returns:
        A list representing the bivector result of the wedge
        ↪  product.
    """
    b12 = v1[0] * v2[1] - v1[1] * v2[0]
    b13 = v1[0] * v2[2] - v1[2] * v2[0]
    b23 = v1[1] * v2[2] - v1[2] * v2[1]
    return [b12, b13, b23]


def product_component(r, a, s, b):
    """
    Computes the geometric product for two homogeneous elements
```

```
    of grades r and s.

    Rules implemented:
      - Grade 0 (scalars): For a scalar a and any component b,
        the product is scalar multiplication and the result retains
        the grade of b.
      - Grade 1 (vectors): For two vectors, the geometric product
        decomposes as:
            a * b = (a·b) + (ab)
        where the dot product yields a grade-0 (scalar) component
        and the wedge product yields a grade-2 (bivector) component.

    Args:
        r: Grade of the first element.
        a: First component.
        s: Grade of the second element.
        b: Second component.

    Returns:
        A dictionary mapping resulting grades to their computed
        ↪    components.

    Raises:
        NotImplementedError for non-supported grade combinations.
    """
    if r == 0:
        # Scalar times any element: result has grade s.
        return {s: scalar_mult_component(a, b)}
    if s == 0:
        # Any element times scalar: result has grade r.
        return {r: scalar_mult_component(b, a)}
    if r == 1 and s == 1:
        # Geometric product of two vectors.
        dp = dot_product(a, b)
        wp = wedge_product(a, b)
        return {0: dp, 2: wp}

    # Other grade combinations are not implemented in this snippet.
    raise NotImplementedError(f"Product for grades {r} and {s} not
    ↪    implemented.")


def geometric_product(mv1, mv2):
    """
    Computes the geometric product of two multivectors.
    It iterates over all pairs of homogeneous components from mv1
    ↪    and mv2,
    applies the corresponding product rule, and sums the results by
    ↪    grade.

    Args:
        mv1: First multivector (dictionary).
        mv2: Second multivector (dictionary).
```

```python
    Returns:
        A multivector (dictionary) representing the geometric
        ↪  product.
    """
    result = {}
    for r, a in mv1.items():
        for s, b in mv2.items():
            try:
                prod = product_component(r, a, s, b)
            except NotImplementedError:
                # Skip combinations which are not implemented.
                continue
            for grade, value in prod.items():
                if grade in result:
                    result[grade] = add_components(result[grade],
                    ↪  value)
                else:
                    result[grade] = value
    return result


if __name__ == "__main__":
    # Demonstration of multivector operations.

    # Create two multivectors with scalar and vector parts.
    mv1 = create_multivector(2, [1, 2, 3])
    mv2 = create_multivector(3, [4, 5, 6])

    print("Multivector 1:", mv1)
    print("Multivector 2:", mv2)

    # Addition of multivectors (grade-wise addition)
    mv_sum = add_multivectors(mv1, mv2)
    print("\nAddition (mv1 + mv2):", mv_sum)

    # Geometric product of multivectors:
    # This includes scalar*scalar, scalar*vector, vector*scalar, and
    ↪  vector*vector parts.
    # For two vectors, the product is decomposed into a scalar (dot
    ↪  product)
    # and a bivector (wedge product) component.
    mv_gp = geometric_product(mv1, mv2)
    print("\nGeometric Product (mv1 * mv2):", mv_gp)

    # Scalar multiplication: multiplying a multivector by a scalar.
    mv_scaled = scalar_multiply_multivector(2, mv1)
    print("\nScalar Multiplication (2 * mv1):", mv_scaled)
```

Chapter 3

Grades and Blades: Understanding the Structure

Grades in Geometric Algebra

Multivectors in geometric algebra are composed of homogeneous components, each characterized by a specific grade. The grade of a component indicates its dimensional nature, with grade 0 representing scalars, grade 1 corresponding to vectors, grade 2 to bivectors, and so on. In a formal mathematical form, any multivector M is decomposed as

$$M = \langle M \rangle_0 + \langle M \rangle_1 + \langle M \rangle_2 + \cdots,$$

where $\langle M \rangle_r$ denotes the homogeneous part of grade r. This systematic decomposition underlies the algebraic structure of geometric algebra by providing a clear hierarchy where each grade encapsulates a distinct subspace. The grading of multivectors facilitates the separation of geometric properties during algebraic manipulations and supports operations that are grade-sensitive.

Blades as Fundamental Components

A blade is defined as a simple, homogeneous multivector that can be expressed as the exterior (wedge) product of a set of linearly independent vectors. For any set of r linearly independent vectors $\{v_1, v_2, \ldots, v_r\}$, the blade is given by

$$B = v_1 \wedge v_2 \wedge \cdots \wedge v_r.$$

This expression encodes the oriented r-dimensional subspace spanned by the vectors. Owing to the antisymmetric properties of the wedge product, blades are inherently decomposable and serve as the basic building blocks for constructing more general multivectors. The simplicity of blades lies in their irreducibility; once a blade is formed by the wedge product, it behaves as an atomic element within its grade. Such properties render blades central to the formulation of transformations and operations—such as rotations, reflections, and projections—in a manner that preserves the geometric information associated with each subspace.

Hierarchical Structure and Computational Representation

The hierarchical nature of geometric algebra arises from the layered composition of multivectors into grades and the further subdivision of each grade into blades. This structure not only organizes the algebraic elements but also enables efficient computational strategies. In numerical implementations, a multivector is commonly represented as a mapping between grades and their corresponding components. This representation supports grade-wise arithmetic and simplifies the algorithmic execution of operations sensitive to the underlying grade.

For instance, the extraction of a particular homogeneous component from a multivector is a fundamental operation in computations. The following Python function illustrates how to retrieve the component of a specified grade from a multivector represented as a dictionary. In this representation, the keys denote the grade, and the associated values correspond to the numerical or vectorial components.

```
def extract_grade(mv, grade):
    """
    Extracts the homogeneous component of a multivector
        corresponding to a specified grade.

    Args:
```

```
    mv: A dictionary representing the multivector, where keys
    ↪    are integer grades.
    grade: The integer grade of interest (e.g., 0 for scalar, 1
    ↪    for vector).

Returns:
    The component associated with the specified grade, or 0 if
    ↪    the grade is not present.
"""
return mv.get(grade, 0)
```

The function above exemplifies a common computational technique, reinforcing the conceptual understanding of grades while streamlining the manipulation of multivector structures. The clear separation between grades, combined with the irreducible nature of blades, forms the backbone of both the theoretical framework and practical implementations within geometric algebra.

Python Code Snippet

```
def extract_grade(mv, grade):
    """
    Extracts the homogeneous component of a multivector
    ↪    corresponding to a specified grade.

    Args:
        mv: A dictionary representing the multivector, where each
        ↪    key is a tuple representing
             a basis blade (an empty tuple for scalars, e.g., () for
        ↪    grade 0; (1,) for e for grade 1;
             (1,2) for a bivector, etc.), and each value is the
        ↪    corresponding coefficient.
        grade: The integer grade of interest.

    Returns:
        A dictionary with only those items whose blade (tuple) has a
        ↪    length equal to the specified grade.
    """
    return {blade: coeff for blade, coeff in mv.items() if
    ↪    len(blade) == grade}


def permutation_sign(lst):
    """
    Computes the sign of the permutation by counting the number of
    ↪    inversions in the list.

    Args:
```

```
        lst: A list of integers representing a permutation of basis
    ↪    indices.

Returns:
    1 if the number of inversions is even (positive sign), or -1
    ↪    if it is odd (negative sign).
"""
inversions = 0
for i in range(len(lst)):
    for j in range(i + 1, len(lst)):
        if lst[i] > lst[j]:
            inversions += 1
return 1 if inversions % 2 == 0 else -1


def wedge_blades(blade1, blade2):
    """
    Computes the wedge (exterior) product of two blades.

    The wedge product of two blades is zero if they share any common
    ↪    basis vectors.
    Otherwise, it produces a new blade (a sorted tuple representing
    ↪    the union of basis indices)
    with an overall sign determined by the permutation required to
    ↪    sort the concatenated list.

    Args:
        blade1: Tuple representing the first blade (e.g., (1,) for
        ↪    e).
        blade2: Tuple representing the second blade (e.g., (2,) for
        ↪    e).

    Returns:
        A tuple (sign, new_blade) if the wedge product is nonzero;
        Returns None if the blades are not disjoint (i.e., their
        ↪    wedge product vanishes).
    """
    # Check if blades share any common basis element.
    if set(blade1).intersection(blade2):
        return None   # The wedge product is zero if there is any
        ↪    overlap.

    # Concatenate the two blades.
    concatenated = list(blade1) + list(blade2)
    # Compute the sorted order of the concatenated list.
    sorted_blade = tuple(sorted(concatenated))
    # Determine the sign factor from the permutation.
    sign = permutation_sign(concatenated)
    return sign, sorted_blade


def wedge_product(mv1, mv2):
    """
```

```
    Computes the wedge (outer) product of two multivectors.

    Each multivector is represented as a dictionary where keys are
    ↪ tuples representing the basis blades
    and values are their coefficients. The wedge product is computed
    ↪ by taking every pair of blades from mv1
    and mv2, applying the wedge product on the blades, and summing
    ↪ up the contributions.

    Args:
        mv1: First multivector (dictionary representation).
        mv2: Second multivector (dictionary representation).

    Returns:
        A dictionary representing the resulting multivector from the
        ↪ wedge product.
    """
    result = {}
    for blade1, coeff1 in mv1.items():
        for blade2, coeff2 in mv2.items():
            res = wedge_blades(blade1, blade2)
            if res is not None:
                sign, new_blade = res
                prod_coeff = coeff1 * coeff2 * sign
                if new_blade in result:
                    result[new_blade] += prod_coeff
                else:
                    result[new_blade] = prod_coeff
    return result


def add_multivectors(mv1, mv2):
    """
    Adds two multivectors represented as dictionaries.

    For each blade present in either multivector, the corresponding
    ↪ coefficients are summed.

    Args:
        mv1: First multivector.
        mv2: Second multivector.

    Returns:
        A dictionary representing the sum of the two multivectors.
    """
    result = mv1.copy()
    for blade, coeff in mv2.items():
        if blade in result:
            result[blade] += coeff
        else:
            result[blade] = coeff
    return result
```

```python
def print_multivector(mv):
    """
    Prints the multivector in a human-readable format.

    The function converts each blade into a string representation.
    For example, the scalar is represented as "1" (empty blade),
    a vector e as "e1", a bivector ee as "e12", etc.

    Args:
        mv: A dictionary representing the multivector.
    """
    if not mv:
        print("0")
        return
    terms = []
    # Sort by grade (length of blade tuple) and lexicographically.
    for blade, coeff in sorted(mv.items(), key=lambda x: (len(x[0]),
    ↪ x[0])):
        if blade == ():
            blade_str = "1"
        else:
            blade_str = "e" + "".join(str(i) for i in blade)
        terms.append(f"{coeff}*{blade_str}")
    print(" + ".join(terms))


# Example usage demonstrating the key equations and algorithmic
↪ ideas from the chapter:
if __name__ == "__main__":
    # Define a multivector M = scalar + vector components.
    # Here, the scalar part is represented with key ().
    # The vector parts: (1,) represents e1 and (2,) represents e2.
    M = {
        (): 3,      # Scalar component 3.
        (1,): 2,    # 2*e1
        (2,): 4     # 4*e2
    }

    print("Multivector M:")
    print_multivector(M)

    # Extract the homogeneous component of grade 1 (vector part).
    grade1 = extract_grade(M, 1)
    print("\nGrade 1 component of M:")
    print_multivector(grade1)

    # Define another multivector N that represents a bivector (a
    ↪ blade)
    # For instance, (1,2) represents e1 e2.
    N = {
        (1, 2): 5   # 5*(e1e2)
    }
```

```
print("\nMultivector N (Bivector):")
print_multivector(N)

# Compute the wedge product: M ∧ N.
# This operation will distribute over the sum in M and combine
↪   blades according to the exterior product.
wedge_M_N = wedge_product(M, N)
print("\nWedge product of M and N (M ∧ N):")
print_multivector(wedge_M_N)

# Demonstrate addition of multivectors: M + N.
sum_M_N = add_multivectors(M, N)
print("\nAddition of M and N (M + N):")
print_multivector(sum_M_N)
```

Chapter 4

The Geometric Product: Definition and Computation

Algebraic Definition of the Geometric Product

The geometric product is defined as a bilinear operation on elements of a Clifford (or geometric) algebra. For any two vectors a and b, the product is given by

$$ab = a \cdot b + a \wedge b,$$

where the symmetric part $a \cdot b$ represents the inner (dot) product and the antisymmetric part $a \wedge b$ corresponds to the exterior (wedge) product. This definition extends by linearity to general multivectors. In a vector space endowed with a Euclidean metric, each basis vector satisfies

$$e_i^2 = 1,$$

and the product of distinct basis vectors adheres to the anticommutation relation

$$e_i e_j = -e_j e_i \quad \text{for } i \neq j.$$

These properties provide an algebraic structure that unifies metric and exterior information in a single product.

Properties and Structural Implications

The geometric product exhibits several important algebraic properties. It is associative,

$$(ab)c = a(bc),$$

and distributive over addition,

$$a(b+c) = ab + ac,$$

which ensures both theoretical elegance and practical utility in computations. Although the product is not commutative in general, the underlying grading leads to useful decomposition rules. In particular, because every multivector M can be decomposed as

$$M = \sum_{r=0}^{n} \langle M \rangle_r,$$

with $\langle M \rangle_r$ denoting the homogeneous component of grade r, some products will preserve or change the grade according to the intrinsic structure of the factors. The symmetric (inner product) and antisymmetric (outer product) components arise naturally from the grading, thus yielding an algebra that captures both metric-dependent and metric-independent information.

It is noteworthy that the noncommutativity of the geometric product encodes geometric relations. In particular, the anticommutative nature of the wedge product component implies that repeated factors cancel in a manner consistent with the geometric interpretation of oriented subspaces. The interplay of these aspects results in a product that is both compact in notation and rich in descriptive power.

Computational Strategies for the Geometric Product

Algorithmic computation of the geometric product is most effectively achieved through the use of basis representations. By expressing multivectors as linear combinations of ordered basis blades, it is possible to exploit the combinatorial structure inherent in the product. In computational implementations, basis blades are commonly represented as sorted tuples of integers denoting the indices of the participating basis vectors. The geometric product of two such basis blades is computed by concatenating the tuples, reordering them into a canonical form, and then eliminating duplicate entries according to the relation $e_i^2 = 1$. The sign of the rearrangement is determined by the parity of the permutation required for sorting.

A fundamental function in this computational framework calculates the geometric product for two basis blades. The function accepts two tuples representing the basis blades. It performs a bubble sort on the concatenated list of indices to deduce the overall sign, and then removes consecutive duplicate indices, which correspond to squared basis vectors. The following Python code snippet illustrates this procedure:

```
def geometric_product_blades(blade1, blade2):
    """
    Computes the geometric product of two basis blades in a
        Euclidean metric space.

    Each basis blade is represented as a tuple of integers
        corresponding to
    the indices of basis vectors. The geometric product is
        calculated by
    concatenating the two blades, reordering them into increasing
        order while
    counting the number of swaps to determine the overall sign.
        Duplicate
    basis indices occurring consecutively are removed based on the
        property
    that e_i^2 = 1.

    Args:
        blade1: Tuple[int, ...]
        blade2: Tuple[int, ...]

    Returns:
        A tuple (sign, result_blade) where sign is either 1 or -1,
            and result_blade
        is the resulting sorted tuple representing the basis blade
            after cancellation.
    """
    combined = list(blade1) + list(blade2)
    swaps = 0
    n = len(combined)
    for i in range(n):
        for j in range(0, n - i - 1):
            if combined[j] > combined[j + 1]:
                combined[j], combined[j + 1] = combined[j + 1],
                    combined[j]
```

```
            swaps += 1
sign = (-1) ** swaps
result = []
i = 0
while i < len(combined):
    if i + 1 < len(combined) and combined[i] == combined[i + 1]:
        i += 2
    else:
        result.append(combined[i])
        i += 1
return sign, tuple(result)
```

The algorithm encapsulated in the function above illustrates the manner in which computational efficiency is achieved. By adhering to the canonical ordering of basis blades and performing the necessary cancellation of duplicate basis indices, the function provides a building block for more elaborate implementations of the geometric product in higher-dimensional spaces.

Python Code Snippet

```
from collections import defaultdict

def geometric_product_blades(blade1, blade2):
    """
    Computes the geometric product of two basis blades in a
    ↪ Euclidean metric space.

    Each basis blade is represented as a tuple of integers
    ↪ corresponding to
    the indices of basis vectors. The geometric product is
    ↪ calculated by
    concatenating the two blades, reordering them into increasing
    ↪ order while
    counting the number of swaps to determine the overall sign.
    ↪ Duplicate
    basis indices occurring consecutively are removed based on the
    ↪ property
    that e_i^2 = 1.

    Args:
        blade1: Tuple[int, ...]
        blade2: Tuple[int, ...]

    Returns:
        A tuple (sign, result_blade) where sign is either 1 or -1,
        ↪ and result_blade
        is the resulting sorted tuple representing the basis blade
        ↪ after cancellation.
```

```
    """
    # Concatenate the two blades
    combined = list(blade1) + list(blade2)
    swaps = 0
    n = len(combined)
    # Bubble sort: reorder combined list and count swaps
    for i in range(n):
        for j in range(0, n - i - 1):
            if combined[j] > combined[j + 1]:
                combined[j], combined[j + 1] = combined[j + 1],
                ↪  combined[j]
                swaps += 1
    sign = (-1) ** swaps
    result = []
    i = 0
    # Remove duplicate indices corresponding e_i^2 = 1
    while i < len(combined):
        if i + 1 < len(combined) and combined[i] == combined[i + 1]:
            i += 2
        else:
            result.append(combined[i])
            i += 1
    return sign, tuple(result)

class Multivector:
    """
    Represents a multivector in geometric algebra.

    A multivector is stored as a dictionary mapping basis blades to
    ↪  their coefficients.
    Basis blades are represented as sorted tuples of integers. The
    ↪  empty tuple () corresponds
    to the scalar part.
    """
    def __init__(self, terms=None):
        # `terms` is a dictionary with keys as basis blades (tuples)
        ↪  and values as coefficients.
        self.terms = {} if terms is None else dict(terms)

    def __add__(self, other):
        """Adds two multivectors."""
        result = Multivector(self.terms)
        for blade, coeff in other.terms.items():
            result.terms[blade] = result.terms.get(blade, 0) + coeff
            if abs(result.terms[blade]) < 1e-12:
                del result.terms[blade]
        return result

    def __sub__(self, other):
        """Subtracts two multivectors."""
        result = Multivector(self.terms)
        for blade, coeff in other.terms.items():
            result.terms[blade] = result.terms.get(blade, 0) - coeff
```

```python
            if abs(result.terms[blade]) < 1e-12:
                del result.terms[blade]
        return result

    def __mul__(self, other):
        """
        Computes the geometric product of two multivectors.

        The product is defined as the sum over the products of
        ↪ individual terms,
        using the geometric product for basis blades.
        """
        result_terms = defaultdict(float)
        for blade1, coeff1 in self.terms.items():
            for blade2, coeff2 in other.terms.items():
                sign, res_blade = geometric_product_blades(blade1,
                ↪ blade2)
                result_terms[res_blade] += coeff1 * coeff2 * sign
        # Filter out near-zero coefficients.
        result = {blade: coeff for blade, coeff in
        ↪ result_terms.items() if abs(coeff) >= 1e-12}
        return Multivector(result)

    def __str__(self):
        """Returns a human-readable string representation of the
        ↪ multivector."""
        if not self.terms:
            return "0"
        terms_str = []
        # Sort terms for a consistent display: first by grade and
        ↪ then lexicographically.
        for blade, coeff in sorted(self.terms.items(), key=lambda x:
        ↪ (len(x[0]), x[0])):
            if blade == ():
                blade_str = "1"
            else:
                blade_str = "".join(f"e{idx}" for idx in blade)
            terms_str.append(f"{coeff:+g}{blade_str}")
        return " ".join(terms_str)

def scalar(value):
    """
    Creates a scalar multivector.

    Args:
        value: numeric value representing the scalar.

    Returns:
        Multivector representing the scalar.
    """
    return Multivector({(): value})

def vector(components):
```

```
"""
Creates a vector multivector.

Args:
    components: dict mapping basis index to coefficient. For
    ↪ example,
                {1: 3, 2: -2} represents the vector 3e1 - 2e2.

Returns:
    Multivector representing the vector.
"""
terms = {}
for idx, coeff in components.items():
    terms[(idx,)] = coeff
return Multivector(terms)

if __name__ == "__main__":
    # Example: Define two vectors a = 2e1 + 3e2 and b = 4e1 - e2
    a = vector({1: 2, 2: 3})
    b = vector({1: 4, 2: -1})

    # Compute the geometric product a*b using multivector
    ↪ multiplication
    product_ab = a * b
    print("Vector a =", a)
    print("Vector b =", b)
    print("Geometric product a * b =", product_ab)

    # Demonstrate the geometric product of two basis blades
    ↪ directly.
    # For instance, e1e2 * e2e3: blade1 = (1, 2), blade2 = (2, 3)
    blade1 = (1, 2)
    blade2 = (2, 3)
    sign, res_blade = geometric_product_blades(blade1, blade2)
    print("\nGeometric product of blades e1e2 and e2e3:")
    print("Sign:", sign, "Resulting blade:", res_blade)

    # Combine multivectors: add a scalar part to the vector.
    mv1 = a + scalar(5)     # Represents 5 + 2e1 + 3e2
    mv2 = b + scalar(-2)    # Represents -2 + 4e1 - e2
    product_mv = mv1 * mv2
    print("\nMultivector mv1 =", mv1)
    print("Multivector mv2 =", mv2)
    print("Geometric product mv1 * mv2 =", product_mv)
```

Chapter 5

Inner Product in GA: Concepts and Formulations

Algebraic Foundations of the Inner Product

In geometric algebra the inner product is introduced as the symmetric portion of the geometric product. For two vectors a and b, the geometric product can be decomposed as

$$ab = a \cdot b + a \wedge b,$$

where the inner product is defined to be

$$a \cdot b = \frac{1}{2}(ab + ba).$$

This definition not only recovers the familiar dot product in a Euclidean space (where each basis vector satisfies $e_i^2 = 1$) but also extends naturally to multivectors of arbitrary grades through a careful extraction of symmetric components. For homogeneous multivectors of grade r and s, the inner product is often defined such that it yields only the component of grade $|r - s|$, thereby preserving the metric properties and encapsulating the projection of one multivector onto another.

Comparative Analysis: Inner Product and Traditional Dot Product

The dot product in traditional vector calculus is a bilinear form that computes the projection of one vector onto another, and in Euclidean spaces it is given by

$$a \cdot b = \sum_{i=1}^{n} a_i b_i.$$

In the setting of geometric algebra, this definition becomes a special case of the inner product when the multivectors are restricted to grade-1 elements. The expression

$$a \cdot b = \frac{1}{2}(ab + ba)$$

demonstrates that the inner product retains the metric-dependent properties of the dot product while simultaneously integrating into a larger algebraic framework that accounts for antisymmetric operations through the wedge product. Such an approach allows the inner product to be generalized to operations between elements of different grades, providing a unified treatment among scalars, vectors, bivectors, and higher-order blades.

Physical Significance of the Inner Product

The inner product is central to many physical applications, offering an efficient mechanism for encoding geometric relationships such as lengths, angles, and projections. In classical mechanics the inner product facilitates the calculation of work or energy through projections of force vectors, while in electromagnetism it underpins the formulation of field interactions. In more advanced theories, including the formulation of spacetime symmetries and Lorentz invariance, the inner product between four-vectors plays a crucial role in preserving invariant quantities. The extraction of the scalar (grade-0) component from the geometric product not only generalizes the notion of distance and orthogonality but also projects the rich structure of the algebra onto physically measurable quantities, thereby bridging abstract algebraic operations with tangible physical observables.

Computational Realization of the Inner Product Operation

From a computational perspective the evaluation of the inner product is achieved by first computing the full geometric product and then isolating its symmetric part. When working with vector representations, the inner product coincides with the traditional dot product and may be efficiently computed by summing the products of corresponding components. A common implementation strategy involves representing vectors as dictionaries that map basis indices to coefficients. In this setting a Python function can be defined to compute the inner product between two vectors. The function operates by traversing the nonzero components of the vector representations and accumulating the product of matching components.

```
def inner_product_vectors(v1, v2):
    """
    Compute the inner product (dot product) of two vectors in a
    ↪ Euclidean geometric algebra.

    Each vector is represented as a dictionary mapping basis index
    ↪ (int)
    to coefficient (float). The inner product is computed as the sum
    ↪ over
    the products of corresponding coefficients, assuming a Euclidean
    ↪ metric
    where basis elements satisfy e_i^2 = 1.

    Args:
        v1: dict[int, float] - vector represented by its components.
        v2: dict[int, float] - vector represented by its components.

    Returns:
        float: The inner product of v1 and v2.
    """
    result = 0.0
    for idx, coeff in v1.items():
        result += coeff * v2.get(idx, 0.0)
    return result
```

This function serves as a computational kernel that embodies the linearity and simplicity of the inner product in the vector space context. Its structure illustrates how abstractions from geometric algebra can be efficiently translated into algorithmic procedures, thereby facilitating numerical simulations and symbolic manipulations in complex physical systems.

Python Code Snippet

```python
def inner_product_vectors(v1, v2):
    """
    Compute the inner product (dot product) of two vectors in a
    ↪ Euclidean geometric algebra.

    Each vector is represented as a dictionary mapping a basis index
    ↪ (an integer)
    to its coefficient (a float). In a Euclidean metric (where e_i^2
    ↪ = 1 for each i),
    the inner (dot) product is given by:

        a · b = sum_{i} a_i * b_i

    Args:
        v1 (dict[int, float]): First vector with keys as indices.
        v2 (dict[int, float]): Second vector with keys as indices.

    Returns:
        float: The computed inner product of v1 and v2.
    """
    result = 0.0
    for idx, coeff in v1.items():
        result += coeff * v2.get(idx, 0.0)
    return result


def wedge_product_vectors(v1, v2):
    """
    Compute the wedge (exterior) product of two vectors, yielding a
    ↪ bivector.

    In geometric algebra the geometric product of two vectors a and
    ↪ b is:

        ab = a·b + a∧b

    where the wedge product a∧b is the antisymmetric part:

        a∧b = 0.5 * (ab - ba)

    For two vectors represented as dictionaries, the bivector is
    ↪ represented as
    another dictionary with keys being tuples (i,j) (with i < j)
    ↪ representing the basis bivector e_i∧e_j.

    The coefficient for the bivector corresponding to the pair (i,j)
    ↪ is given by:

        (a_i * b_j) - (a_j * b_i)
```

```
    Note: Components where i == j do not contribute since they
    ↪    cancel out.

    Args:
        v1 (dict[int, float]): First vector.
        v2 (dict[int, float]): Second vector.

    Returns:
        dict[tuple[int, int], float]: A dictionary representing the
        ↪    bivector components.
    """
    bivector = {}
    for i, coeff1 in v1.items():
        for j, coeff2 in v2.items():
            if i < j:
                bivector[(i, j)] = bivector.get((i, j), 0.0) +
                ↪    coeff1 * coeff2
            elif i > j:
                # When i > j, switch the order and add with a
                ↪    negative sign (antisymmetry).
                bivector[(j, i)] = bivector.get((j, i), 0.0) -
                ↪    coeff1 * coeff2
            # When i == j, the contribution is part of the symmetric
            ↪    (inner) product.
    return bivector


def geometric_product_vectors(v1, v2):
    """
    Compute the full geometric product of two vectors in a Euclidean
    ↪    geometric algebra.

    For two vectors a and b, the geometric product is the sum of:

        a · b      (symmetric inner part, grade-0)
        a ∧ b      (antisymmetric outer part, grade-2 aka bivector)

    This function returns a dictionary with two entries:

        'scalar': The inner/dot product result (float).
        'bivector': The wedge product result (dict with tuple keys
        ↪    representing the bivector basis elements).

    Args:
        v1 (dict[int, float]): First vector.
        v2 (dict[int, float]): Second vector.

    Returns:
        dict: A dictionary with keys 'scalar' and 'bivector'
        ↪    representing the multivector.
    """
    scalar_part = inner_product_vectors(v1, v2)
    bivector_part = wedge_product_vectors(v1, v2)
```

```python
    return {"scalar": scalar_part, "bivector": bivector_part}


if __name__ == '__main__':
    # Example vectors represented in the Euclidean space basis.
    # For instance, let:
    #   v1 = 3e1 + 2e2   represented as   {1: 3.0, 2: 2.0}
    #   v2 = 1e1 + 4e2   represented as   {1: 1.0, 2: 4.0}

    v1 = {1: 3.0, 2: 2.0}
    v2 = {1: 1.0, 2: 4.0}

    # Compute the inner product (dot product)
    dot_product = inner_product_vectors(v1, v2)
    print("Inner Product (a · b):", dot_product)

    # Compute the wedge product (bivector part)
    bivector = wedge_product_vectors(v1, v2)
    print("Wedge Product (a  b):", bivector)

    # Compute the full geometric product (scalar + bivector)
    geo_product = geometric_product_vectors(v1, v2)
    print("Geometric Product (ab):", geo_product)

    # The output dictionary 'geo_product' shows:
    #   - 'scalar': the inner product component.
    #   - 'bivector': the bivector components, representing the
    #     oriented plane segment.
```

Chapter 6

Outer Product: The Wedge Product Essentials

Geometric Interpretation of the Wedge Product

The exterior product, denoted by

$$a \wedge b,$$

is defined as the antisymmetric portion of the geometric product. For any two vectors a and b, the decomposition

$$ab = a \cdot b + a \wedge b$$

implies that the wedge product captures the oriented area of the parallelogram spanned by a and b. Its antisymmetric nature is evident from the identity

$$a \wedge b = -b \wedge a,$$

which immediately ensures that for any vector v

$$v \wedge v = 0.$$

The magnitude of $a \wedge b$ corresponds to the area determined by the two vectors, while its sign encodes the orientation of the plane defined by them. In higher-dimensional spaces, the repeated application of the wedge product constructs multivectors of increasing grade, where a product such as

$$v_1 \wedge v_2 \wedge \cdots \wedge v_r$$

defines an oriented r-dimensional volume element. This elegant geometric interpretation unifies concepts of area, volume, and hypervolume within a single algebraic framework.

Algebraic Structure and Construction of Higher-Grade Elements

The wedge product is central to the construction of blades, which are multivectors that represent subspaces with a well-defined orientation and magnitude. Given a set of r linearly independent vectors, the blade

$$B = v_1 \wedge v_2 \wedge \cdots \wedge v_r$$

uniquely characterizes the r-dimensional subspace spanned by these vectors. The antisymmetry of the wedge product guarantees that any repetition of vectors results in a null element, ensuring that only independent directions contribute to the grade of B. Algebraically, the operation is both bilinear and associative, which permits the systematic assembly of higher-grade elements from lower-grade components.

When the geometric product of two vectors is expanded, the inner product isolates the scalar (grade-0) portion while the wedge product extracts the bivector (grade-2) component. This separation is formally given by

$$a \wedge b = \frac{1}{2}(ab - ba).$$

The algebraic consistency provided by this structure allows for the extension of these operations to multivectors of arbitrary grade, thereby facilitating the modeling of complex geometric and physical systems.

Computational Implementation of the Wedge Product

Computations within the geometric algebra framework frequently employ sparse representations of vectors, where each vector is stored as a mapping from basis indices to numerical coefficients. In this computational context, the wedge product of two vectors is evaluated by iterating over pairs of indices from the union of the bases and computing contributions only for those pairs satisfying $i < j$. The antisymmetry of the wedge product mandates that each pair (i, j) yields a contribution of the form

$$v_1(i)\, v_2(j) - v_1(j)\, v_2(i).$$

A single function that encapsulates this procedure is provided below. This function assumes that the vectors are represented as dictionaries mapping each index to its corresponding coefficient and returns a dictionary representing the bivector components. Each key in the resulting dictionary is a tuple (i, j) with $i < j$, and the associated value is the corresponding wedge product coefficient.

```
def wedge_product_vectors(v1, v2):
    """
    Compute the wedge (exterior) product of two vectors in a
    ↪ Euclidean
    geometric algebra.

    The wedge product for vectors a and b is given by:
        a  b = a_i * b_j - a_j * b_i,
    where the computation is over all pairs of basis indices (i, j)
    ↪ with i < j.
    In this representation, each vector is a dictionary mapping
    ↪ basis index (int)
    to its coefficient (float), and the result is a dictionary
    ↪ mapping the tuple (i, j)
    to the computed bivector coefficient corresponding to the basis
    ↪ element e_i  e_j.

    Args:
        v1 (dict[int, float]): The first vector represented in
        ↪ sparse form.
        v2 (dict[int, float]): The second vector represented in
        ↪ sparse form.

    Returns:
        dict[tuple[int, int], float]: A dictionary with keys as
        ↪ tuples (i, j) where i < j,
```

```
        and values as the corresponding wedge product coefficients.
    """
    bivector = {}
    # Define the sorted union of indices from both vectors.
    indices = sorted(set(v1.keys()).union(v2.keys()))
    for idx, i in enumerate(indices):
        for j in indices[idx+1:]:
            coeff = v1.get(i, 0.0) * v2.get(j, 0.0) - v1.get(j, 0.0)
            ↪    * v2.get(i, 0.0)
            if coeff != 0:
                bivector[(i, j)] = coeff
    return bivector
```

Python Code Snippet

```
# Outer Product and Related Operations in Geometric Algebra
#
# This module implements key operations from the chapter,
# including the wedge (exterior) product for vectors, the inner
    ↪ (dot)
# product, the full geometric product (decomposed into scalar and
    ↪ bivector parts),
# and the recursive computation of a wedge product over an arbitrary
    ↪ number of vectors.
# Vectors are represented in sparse form as dictionaries mapping
    ↪ basis indices (int)
# to their coefficients (float). Multivectors (blades) are similarly
    ↪ represented,
# with keys as tuples of basis indices (in sorted order) and values
    ↪ as the corresponding coefficients.

def wedge_product_vectors(v1, v2):
    """
    Compute the wedge (exterior) product of two vectors in a
        ↪ Euclidean geometric algebra.

    For two vectors a and b represented as:
        a = a[i] e_i,    and   b = b[i] e_i,
    the wedge product is defined as:
        a ∧ b = ∑_{i<j} (a[i]*b[j] - a[j]*b[i]) (e_i ∧ e_j),
    which extracts the oriented area (a bivector) spanned by a and
        ↪ b.

    Args:
        v1 (dict[int, float]): First vector in sparse form.
        v2 (dict[int, float]): Second vector in sparse form.

    Returns:
        dict[tuple[int, int], float]: Dictionary of bivector
            ↪ components.
```

```
            Each key is a tuple (i, j) (with i < j) representing the
            ↪ basis element e_i  e_j.
    """
    bivector = {}
    # Form the sorted union of basis indices from both vectors.
    indices = sorted(set(v1.keys()).union(v2.keys()))
    for idx, i in enumerate(indices):
        for j in indices[idx+1:]:
            coeff = v1.get(i, 0.0) * v2.get(j, 0.0) - v1.get(j, 0.0)
            ↪ * v2.get(i, 0.0)
            if coeff != 0.0:
                bivector[(i, j)] = coeff
    return bivector

def inner_product_vectors(v1, v2):
    """
    Compute the inner (dot) product of two vectors in a Euclidean
    ↪ space.

    The dot product is given by:
        a · b = _{i} a[i]*b[i],
    summing over the indices common to both vectors.

    Args:
        v1 (dict[int, float]): First vector.
        v2 (dict[int, float]): Second vector.

    Returns:
        float: The scalar (grade-0) result of the inner product.
    """
    common_indices = set(v1.keys()).intersection(v2.keys())
    total = sum(v1[i] * v2[i] for i in common_indices)
    return total

def geometric_product_vectors(v1, v2):
    """
    Compute the full geometric product of two vectors.

    In geometric algebra, the geometric product is decomposed as:
        ab = a · b + a  b,
    where a · b (the inner product) produces the scalar part,
    and a  b (the wedge product) produces the bivector part.

    Args:
        v1 (dict[int, float]): First vector.
        v2 (dict[int, float]): Second vector.

    Returns:
        tuple[float, dict[tuple[int, int], float]]:
            - The scalar component (inner product).
            - The bivector component (wedge product), represented as
              ↪ a dictionary.
    """
```

```python
        inner = inner_product_vectors(v1, v2)
        outer = wedge_product_vectors(v1, v2)
        return inner, outer

def wedge_product_n(*vectors):
    """
    Compute the wedge product of an arbitrary number of vectors,
    constructing a higher-grade blade representing an oriented
    ↪ subspace.

    The wedge product is computed recursively. A vector is first
    ↪ cast into a
    blade with one-element tuple keys. Then, for each additional
    ↪ vector, the wedge
    product with the current blade is computed. The antisymmetry is
    ↪ handled by:
      - Skipping duplicate basis elements.
      - Sorting the resulting multi-index and adjusting the sign
      ↪ according to the
        number of transpositions required.

    Args:
        *vectors: A variable number of vectors (each as a dict[int,
        ↪ float]).

    Returns:
        dict: A dictionary representing the multivector blade.
              Keys are tuples of basis indices (sorted in increasing
              ↪ order),
              and values are the coefficients.
    """
    if not vectors:
        return {}

    # Initialize the blade with the first vector.
    # Represent each basis element as a one-tuple.
    blade = { (i,): coeff for i, coeff in vectors[0].items() }

    def wedge_blade_vector(blade, v):
        """
        Compute the wedge product of an existing blade with a
        ↪ vector.

        Args:
            blade (dict[tuple, float]): The current blade
            ↪ (multivector).
            v (dict[int, float]): The vector to wedge with.

        Returns:
            dict: The updated blade after wedging with the vector.
        """
        new_blade = {}
        for basis, coeff in blade.items():
```

```python
            for i, val in v.items():
                if i in basis:
                    # Antisymmetry: wedge product with a repeated
                    ↪  vector gives zero.
                    continue
                # Form the new multi-index by appending and then
                ↪  sorting.
                new_basis = list(basis) + [i]
                sorted_basis = sorted(new_basis)
                # Determine the sign of the permutation from
                ↪  new_basis to sorted_basis.
                sign = 1
                temp_basis = list(new_basis)
                for j in range(len(temp_basis)):
                    for k in range(j+1, len(temp_basis)):
                        if temp_basis[j] > temp_basis[k]:
                            sign *= -1
                new_key = tuple(sorted_basis)
                new_blade[new_key] = new_blade.get(new_key, 0) +
                ↪  coeff * val * sign
        return new_blade

    # Recursively compute the wedge product with remaining vectors.
    for v in vectors[1:]:
        blade = wedge_blade_vector(blade, v)
    return blade

# Example usage and demonstration of the implemented functions.
if __name__ == "__main__":
    # Define two example vectors in 3-dimensional space.
    # Here, indices 0, 1, and 2 correspond to basis elements e1, e2,
    ↪  and e3, respectively.
    v1 = {0: 3.0, 1: 4.0}          # Represents 3e1 + 4e2.
    v2 = {0: 1.0, 1: -2.0, 2: 5.0}  # Represents 1e1 - 2e2 + 5e3.

    print("Vector v1:", v1)
    print("Vector v2:", v2)

    # Compute the wedge product (bivector) of v1 and v2.
    bivector = wedge_product_vectors(v1, v2)
    print("\nWedge Product (v1  v2):", bivector)

    # Compute the inner (dot) product of v1 and v2.
    inner = inner_product_vectors(v1, v2)
    print("\nInner Product (v1 · v2):", inner)

    # Compute the full geometric product of v1 and v2.
    geo_inner, geo_outer = geometric_product_vectors(v1, v2)
    print("\nGeometric Product (v1 * v2):")
    print("  Scalar (inner product):", geo_inner)
    print("  Bivector (wedge product):", geo_outer)
```

```python
# Define a third vector for computing a 3-blade (oriented volume
↪   element).
v3 = {1: 2.0, 2: -3.0}   # Represents 2e2 - 3e3.
print("\nVector v3:", v3)

# Compute the wedge product of v1, v2, and v3.
blade_3 = wedge_product_n(v1, v2, v3)
print("\nWedge Product (v1   v2   v3):", blade_3)
```

Chapter 7

Clifford Algebras: Foundations for Computational Physics

Mathematical Foundations

A Clifford algebra is constructed from a real vector space V equipped with a quadratic form Q. The algebra $\mathcal{C}\ell(V, Q)$ is defined as the associative algebra generated by the elements of V subject to the relation
$$v^2 = Q(v) \quad \text{for all } v \in V.$$
In the context of an orthonormal basis $\{e_i\}$, this defining relation takes the form
$$e_i e_j + e_j e_i = 2g_{ij},$$
where g_{ij} are the elements of the metric tensor with $g_{ii} = Q(e_i)$ and $g_{ij} = 0$ for $i \neq j$. This relation not only generalizes the familiar inner product but also unifies it with the exterior (wedge) product within a single algebraic structure.

Algebraic Structure and Product Operations

The elements of a Clifford algebra can be expressed as multivectors, which are linear combinations of basis blades. A basis blade is formed by the ordered Clifford product of a set of generators. For indices i and j with $i \neq j$, the antisymmetry inherent in the generators yields

$$e_i\, e_j = -e_j\, e_i.$$

A general Clifford product between two basis blades e_I and e_J, where I and J are ordered tuples of indices, can be expressed symbolically as

$$e_I\, e_J = \epsilon(I, J)\, e_{I \oplus J},$$

where $\epsilon(I, J)$ is a sign factor determined by the number of transpositions necessary to bring the concatenated index tuple $I \cup J$ into its canonical order, and the operator $\oplus$ indicates the contraction of identical indices via the metric. When a repeated index is encountered, the relation

$$e_i^2 = g_{ii}$$

applies, effecting a reduction of the product and a scalar contribution from the metric.

Computational Representation and Algorithmic Implementation

In computational applications, basis blades are efficiently represented as sorted tuples of integers corresponding to generator indices. The Clifford product is then implemented by concatenating the tuples representing two basis blades, reordering them into canonical form, and tracking sign changes due to the required permutations. Duplicate indices are removed by invoking the corresponding metric relation, thereby introducing the appropriate scalar multiplication.

The following function exemplifies this procedure by computing the Clifford product for two basis blades. Given the blades represented as sorted tuples and a dictionary representing the signature of the metric, the function returns a scalar coefficient along with the resulting blade in its canonical form.

```
def clifford_product_basis(b1, b2, metric):
    """
    Compute the Clifford product of two basis blades in a Clifford
    ↪ algebra.
```

```
Parameters:
    b1 (tuple[int]): A sorted tuple representing the first basis
    ↪   blade.
    b2 (tuple[int]): A sorted tuple representing the second
    ↪   basis blade.
    metric (dict[int, float]): A dictionary mapping basis
    ↪   indices to their quadratic form values.
                        Typically, for an orthonormal
                        ↪   basis, g[i] is either 1 or
                        ↪   -1.

Returns:
    tuple[float, tuple[int]]: A tuple (coeff, blade) where
    ↪   'coeff' is the scalar factor
                        resulting from the reordering and
                        ↪   metric contractions, and
                        'blade' is the sorted tuple
                        ↪   representing the resulting
                        ↪   basis blade.
"""
# Concatenate the indices of both blades.
combined = list(b1) + list(b2)

# Determine the sign factor by counting transpositions using a
↪   bubble sort.
transpositions = 0
arr = combined[:]
n = len(arr)
for i in range(n):
    for j in range(0, n - i - 1):
        if arr[j] > arr[j + 1]:
            arr[j], arr[j + 1] = arr[j + 1], arr[j]
            transpositions += 1
sign = (-1) ** transpositions
sorted_indices = arr

# Reduce duplicate indices via the metric.
result_blade = []
for index in sorted_indices:
    if result_blade and result_blade[-1] == index:
        result_blade.pop()
        sign *= metric[index]
    else:
        result_blade.append(index)
return sign, tuple(result_blade)
```

In the above function, the two input blades e_I and e_J are provided as sorted tuples. The concatenation of their indices yields a temporary list that is subsequently sorted; the number of swaps performed in this process determines the overall sign factor $(-1)^{\text{transpositions}}$. The sorting not only imposes the canonical order but also facilitates the detection and removal of duplicate indices. Each occurrence of a duplicate, which corresponds to a squared generator, is removed from the blade with the sign being adjusted by the metric value of that generator. The resulting tuple represents the canonical form of the product, and the computed sign encapsulates both permutation and metric contributions.

Relevance in Modeling Physical Phenomena

Clifford algebras provide a robust framework for describing geometric transformations and physical interactions. They naturally incorporate rotations, reflections, and Lorentz transformations by facilitating the algebraic manipulation of multivectors. For example, a vector $v = \sum_i v_i \, e_i$ in a physical space obeys the relation

$$v^2 = \sum_i v_i^2 \, g_{ii},$$

which reflects the underlying quadratic form. The algebraic operations of a Clifford algebra, in particular the Clifford product, allow for a unified treatment of scalars, vectors, and higher-grade elements, thereby enabling comprehensive models for phenomena such as electromagnetism, quantum spin, and relativistic kinematics.

The ability to represent these complex interactions in a computationally tractable manner has significant implications. By encoding the non-commutative properties of the basis elements and incorporating metric information directly into the product operation, Clifford algebras serve as the foundation for numerous computational algorithms in physics. The seamless merger of geometric intuition with algorithmic precision renders Clifford algebras indispensable in computational physics, where the accurate simulation of physical systems often relies on the inherent structure of the algebra.

Python Code Snippet

```
def clifford_product_basis(b1, b2, metric):
    """
```

```
    Compute the Clifford product of two basis blades in a Clifford
 ↪   algebra.

    Parameters:
        b1 (tuple[int]): A sorted tuple representing the first basis
 ↪       blade.
        b2 (tuple[int]): A sorted tuple representing the second
 ↪       basis blade.
        metric (dict[int, float]): A dictionary mapping basis
 ↪       indices to their quadratic form values.
                                   For an orthonormal basis,
                                    ↪  metric[i] is typically 1 or
                                    ↪  -1.

    Returns:
        tuple[float, tuple[int]]: A tuple (coeff, blade) where
 ↪       'coeff' is the scalar factor
                                    resulting from the reordering and
                                    ↪  metric contractions, and
                                    'blade' is the sorted tuple
                                    ↪  representing the resulting
                                    ↪  basis blade.
    """
    # Concatenate the indices from both blades.
    combined = list(b1) + list(b2)

    # Sort the indices using a simple bubble sort while counting
 ↪   transpositions for the sign.
    transpositions = 0
    arr = combined[:]    # Create a copy of the list for sorting.
    n = len(arr)
    for i in range(n):
        for j in range(0, n - i - 1):
            if arr[j] > arr[j + 1]:
                arr[j], arr[j + 1] = arr[j + 1], arr[j]
                transpositions += 1
    sign = (-1) ** transpositions
    sorted_indices = arr

    # Reduce duplicate indices using the metric relation: e_i^2 =
 ↪   g_{ii}.
    result_blade = []
    for index in sorted_indices:
        if result_blade and result_blade[-1] == index:
            result_blade.pop()
            sign *= metric[index]
        else:
            result_blade.append(index)
    return sign, tuple(result_blade)


def multiply_multivectors(M1, M2, metric):
    """
```

```
    Multiply two multivectors represented as dictionaries mapping
    ↪ basis blades to coefficients.

    Parameters:
        M1 (dict[tuple[int], float]): The first multivector.
        M2 (dict[tuple[int], float]): The second multivector.
        metric (dict[int, float]): The metric dictionary for the
            ↪ Clifford algebra.

    Returns:
        dict[tuple[int], float]: The resulting multivector after
            ↪ multiplication.
    """
    result = {}
    # Loop over each term (blade and its coefficient) in the first
        ↪ multivector.
    for blade1, coeff1 in M1.items():
        # Loop over each term in the second multivector.
        for blade2, coeff2 in M2.items():
            # Compute the Clifford product for the basis blades.
            sign, res_blade = clifford_product_basis(blade1, blade2,
                ↪ metric)
            new_coeff = coeff1 * coeff2 * sign
            # Accumulate the coefficient for identical resulting
                ↪ blades.
            if res_blade in result:
                result[res_blade] += new_coeff
            else:
                result[res_blade] = new_coeff
    return result


def display_multivector(M):
    """
    Display the multivector in a human-readable format.

    Parameters:
        M (dict[tuple[int], float]): The multivector to display.
    """
    terms = []
    # Sort the blades by grade (length of tuple) and then
        ↪ lexicographically.
    for blade, coeff in sorted(M.items(), key=lambda x: (len(x[0]),
        ↪ x[0])):
        if abs(coeff) < 1e-10:  # Skip near-zero coefficients.
            continue
        if blade == ():
            term = f"{coeff:.2f}"
        else:
            # Construct a string representation like 'e1e2' for the
                ↪ blade.
            generators = "".join([f"e{idx}" for idx in blade])
            term = f"{coeff:.2f}{generators}"
```

```python
        terms.append(term)
    print(" + ".join(terms) if terms else "0")


if __name__ == "__main__":
    # Define the metric for a 3D Euclidean space (orthonormal basis)
    ↪ where e_i^2 = 1.
    metric = {1: 1, 2: 1, 3: 1}

    # Example: Compute the Clifford product of two basis blades.
    blade1 = (1, 2)  # Represents e1e2.
    blade2 = (2, 3)  # Represents e2e3.
    coeff, res_blade = clifford_product_basis(blade1, blade2,
    ↪ metric)
    print("Clifford product of", blade1, "and", blade2, "is:",
    ↪ coeff, res_blade)

    # Define two multivectors as dictionaries.
    # The key is a tuple representing the basis blade (empty tuple
    ↪ () represents the scalar part)
    # and the value is the coefficient.
    # For example, let M = 3 + 2e1 + 0.5e1e2 and N = 1 - e2 + e2e3.
    M = {(): 3, (1,): 2, (1, 2): 0.5}
    N = {(): 1, (2,): -1, (2, 3): 1}

    print("\nMultivector M:")
    display_multivector(M)
    print("Multivector N:")
    display_multivector(N)

    # Compute the product of the multivectors M and N.
    product = multiply_multivectors(M, N, metric)
    print("\nProduct of M and N:")
    display_multivector(product)
```

Chapter 8

Basis and Representation: Constructing the Algebra

Basis Choices in Geometric Algebra

The fundamental structure of geometric algebra is established by selecting an appropriate basis for the underlying vector space V. Often, an orthonormal basis $\{e_1, e_2, \ldots, e_n\}$ is chosen so that each generator satisfies

$$e_i^2 = \pm 1,$$

with the sign dictated by the metric signature. Alternative bases, such as null or non-orthogonal systems, may be employed according to the symmetry properties or the physical context under study. In each case, the canonical ordering of the basis elements is of paramount importance, as it governs the algebraic manipulation of blades—products of basis vectors that are antisymmetric under exchange. Blades of increasing grade are constructed by the wedge (exterior) product of these elements, and the resulting sign changes caused by index permutations are critical to maintain consistent transformation laws throughout the computation. This systematic organization underpins both theoretical analysis and algorithmic implementation.

Coordinate Representations and Computational Implications

Translating the abstract algebraic entities into concrete computational objects involves assigning coordinates to each basis element. A vector $v \in V$ is typically expressed as

$$v = \sum_i v^i\, e_i,$$

where the coefficients v^i represent its components with respect to the chosen basis. This expansion naturally extends to multivectors, which are linear combinations of blades spanning grades from 0 (scalars) up to n. In computational practice, each multivector is stored as a set of nonzero terms, with each term associated with a unique basis blade.

A common strategy is to represent basis blades in their canonical form as sorted tuples of integers. This representation encodes the algebra's combinatorial structure and facilitates the implementation of operations such as the Clifford product. The metric tensor, integrated into the coordinate representation, determines the behavior of the product operations by dictating how basis elements combine under squared operations and contractions. This approach provides an effective means to transition from the continuous mathematics of geometric algebra to discrete numerical algorithms.

Explicit Construction of Multivectors for Computational Tasks

In computational applications, a multivector is constructed explicitly as a linear combination of basis blades, each of which is uniquely identified by its canonical representation. Typically, a multivector is implemented as a dictionary where the keys are sorted tuples of integers corresponding to basis blades, and the associated values are the scalar coefficients. This sparse representation is highly efficient, given that many physical systems yield multivectors with a limited number of nonzero terms.

A representative function that builds a multivector from a list of terms is provided below. Each term consists of a basis blade—represented as a tuple of integers—and its coefficient. The function aggregates coefficients for blades that appear multiple times, ensuring the resulting multivector conforms to the canonical structure required for subsequent algebraic operations.

```
def build_multivector(terms):
    """
    Construct a multivector from a list of terms.

    Parameters:
        terms (list[tuple[tuple[int], float]]): A list where each
          element is a tuple
            consisting of a basis blade (represented as a tuple of
          integers) and its
            corresponding coefficient.

    Returns:
        dict[tuple[int], float]: A dictionary representing the
          multivector, where keys are
            basis blades in their canonical form and values are the
          summed coefficients.
    """
    multivector = {}
    for blade, coeff in terms:
        if blade in multivector:
            multivector[blade] += coeff
        else:
            multivector[blade] = coeff
    return multivector
```

Through this explicit construction, each basis blade is managed in a manner that facilitates subsequent operations, such as computing the geometric product or applying linear transformations. The ordered tuples ensure consistent handling of permutation-induced sign factors, while the dictionary structure enables rapid lookup and efficient manipulation of nonzero components.

In computational frameworks, the interplay between basis selection, coordinate assignment, and explicit multivector construction forms a robust foundation. This framework permits the precise translation of abstract algebraic notions into algorithms capable of simulating complex physical phenomena with high fidelity.

Python Code Snippet

```
# Python code snippet for constructing and manipulating multivectors
  in Geometric Algebra
```

```python
def build_multivector(terms):
    """
    Construct a multivector from a list of terms.

    Parameters:
        terms (list[tuple[tuple[int], float]]): A list where each
        ↪ element is a tuple
            consisting of a basis blade (represented as a tuple of
            ↪ integers) and its
            corresponding coefficient.

    Returns:
        dict[tuple[int], float]: A dictionary representing the
        ↪ multivector, where keys are
            basis blades in their canonical form and values are the
            ↪ summed coefficients.
    """
    multivector = {}
    for blade, coeff in terms:
        if blade in multivector:
            multivector[blade] += coeff
        else:
            multivector[blade] = coeff
    return multivector


def canonical_blade(blade):
    """
    Return the canonical form (sorted tuple) of the blade along with
    ↪ its sign factor.

    The sign factor accounts for the number of swaps needed to sort
    ↪ the blade.

    Parameters:
        blade (tuple[int]): A tuple representing the indices of a
        ↪ basis blade.

    Returns:
        tuple: (canonical_blade, sign) where canonical_blade is a
        ↪ sorted tuple and sign is +1 or -1.
    """
    blade_list = list(blade)
    sign = 1
    # Simple bubble-sort algorithm to count the swaps (permutation
    ↪ parity)
    for i in range(len(blade_list)):
        for j in range(i + 1, len(blade_list)):
            if blade_list[i] > blade_list[j]:
                blade_list[i], blade_list[j] = blade_list[j],
                ↪ blade_list[i]
                sign = -sign
    return tuple(blade_list), sign
```

```python
def multiply_blades(blade1, blade2, metric):
    """
    Compute the geometric product of two basis blades.

    The algorithm performs the following:
      1. Concatenates the indices of blade1 and blade2.
      2. Sorts the resulting list to obtain the canonical form,
         tracking the sign factor.
      3. Detects duplicate indices (which indicate squared basis
         elements) and removes them,
           multiplying in the corresponding metric factor.

    Parameters:
        blade1 (tuple[int]): The first basis blade.
        blade2 (tuple[int]): The second basis blade.
        metric (dict): A dictionary mapping basis indices to their
           square (e.g., {1: 1, 2: 1, ...}).

    Returns:
        tuple: (result_blade, overall_factor) where result_blade is
           the canonical basis blade
             (with duplicate indices removed) and overall_factor
                is the product of the sign and metric factors.
    """
    # Concatenate the two blades
    combined = list(blade1) + list(blade2)

    # Obtain canonical form and sign factor
    canonical, sign = canonical_blade(combined)

    # Remove duplicate indices while applying the metric factors
    result = []
    overall_factor = sign
    i = 0
    while i < len(canonical):
        if i + 1 < len(canonical) and canonical[i] == canonical[i +
           1]:
            # Duplicate found: multiply by the square of the basis
               vector (e_i^2) from the metric
            idx = canonical[i]
            overall_factor *= metric.get(idx, 1)
            i += 2  # Skip the next duplicate as well
        else:
            result.append(canonical[i])
            i += 1
    return tuple(result), overall_factor


def geometric_product(mv1, mv2, metric):
    """
    Compute the geometric product of two multivectors.
```

*Each multivector is represented as a dictionary where the keys
    are basis blades (tuples)
and the values are their corresponding scalar coefficients.*

*The product is computed by multiplying every term in mv1 with
    every term in mv2,
using the basis blade multiplication defined in multiply_blades,
    and then aggregating
like terms.*

*Parameters:
    mv1 (dict): The first multivector.
    mv2 (dict): The second multivector.
    metric (dict): The metric defining basis vector squares.*

*Returns:
    dict: A dictionary representing the resulting multivector
        after the geometric product.*
"""
```
result = {}
for blade1, coeff1 in mv1.items():
    for blade2, coeff2 in mv2.items():
        prod_blade, factor = multiply_blades(blade1, blade2,
            metric)
        new_coeff = coeff1 * coeff2 * factor
        if prod_blade in result:
            result[prod_blade] += new_coeff
        else:
            result[prod_blade] = new_coeff
# Filter out terms with coefficients near zero
result = {blade: coeff for blade, coeff in result.items() if
    abs(coeff) > 1e-12}
return result


def display_multivector(mv):
    """
```
Produce a human-readable string representation of a multivector.

*Parameters:
    mv (dict): A multivector represented as a dictionary with
        basis blade tuples as keys.*

*Returns:
    str: A string representation of the multivector.*
"""
```
terms = []
# Sort the blades by grade (the length of the tuple) and
    lexicographically.
for blade, coeff in sorted(mv.items(), key=lambda x: (len(x[0]),
    x[0])):
    if blade == ():
```

```
            terms.append(f"{coeff}")
        else:
            basis_str = "".join(f"e{idx}" for idx in blade)
            terms.append(f"{coeff}*{basis_str}")
    return " + ".join(terms) if terms else "0"


if __name__ == "__main__":
    # Define a metric for a 3-dimensional Euclidean space.
    # In Euclidean space, each basis vector squares to 1.
    metric = {1: 1, 2: 1, 3: 1}

    # Construct two example multivectors.
    # Example multivector mv1: v = 3*e1 + 2*e2
    mv1 = build_multivector([((1,), 3.0), ((2,), 2.0)])

    # Example multivector mv2: u = 4*e1 + 1*e2 + 5 (scalar)
    mv2 = build_multivector([((1,), 4.0), ((2,), 1.0), ((), 5.0)])

    # Compute the geometric product of mv1 and mv2
    gp = geometric_product(mv1, mv2, metric)

    # Display the input multivectors and their geometric product.
    print("Multivector 1: ", display_multivector(mv1))
    print("Multivector 2: ", display_multivector(mv2))
    print("Geometric Product: ", display_multivector(gp))
```

Chapter 9

Reversion Operation: Properties and Computation

Mathematical Definition of Reversion

In geometric algebra, every multivector M can be decomposed into components of different grades:

$$M = \sum_{k=0}^{n} \langle M \rangle_k,$$

where $\langle M \rangle_k$ denotes the homogeneous grade-k part. The reversion (or reversal) operation is defined by reversing the order of the basis vectors composing each blade. For a grade-k element, the reversal is given by

$$\widetilde{\langle M \rangle_k} = (-1)^{\frac{k(k-1)}{2}} \langle M \rangle_k.$$

This formula captures the effect that reversing the order of a product of k basis vectors introduces a sign change determined by the combinatorial factor $\frac{k(k-1)}{2}$. For instance, scalar elements (grade 0) remain unchanged since

$$(-1)^{\frac{0(0-1)}{2}} = 1,$$

and vectors (grade 1) also remain invariant

$$(-1)^{\frac{1(1-1)}{2}} = 1.$$

Bivectors (grade 2) acquire a negative sign as

$$(-1)^{\frac{2(2-1)}{2}} = -1,$$

and similar patterns hold for higher-grade elements. This operation is an involution; that is, applying the reversion twice returns the original multivector:

$$\widetilde{(\widetilde{M})} = M.$$

Properties of the Reversion Operation

The reversion operation exhibits several key algebraic properties that are essential within geometric algebra. It acts as an anti-automorphism over the algebra, so for any two multivectors A and B,

$$\widetilde{AB} = \widetilde{B}\,\widetilde{A}.$$

This property plays a significant role in simplifying expressions, particularly when dealing with products of multivectors.

Linearity is maintained through the operation, and the reversion preserves the scalar part while multiplying higher-grade elements by the appropriate factor as dictated by the formula

$$(-1)^{\frac{k(k-1)}{2}},$$

where k is the grade. Unlike other involutions in geometric algebra, such as the grade involution or Clifford conjugation, the reversion specifically targets the ordering of basis vectors, ensuring that the antisymmetric structure of blades is rigorously respected.

Computational Implementation of Reversion

Efficient computational implementation of the reversion operation relies on taking advantage of the structured representation of multivectors. In many computational approaches, a multivector is stored as a dictionary where each key is a canonical basis blade represented as a sorted tuple of integers and the associated value is the scalar coefficient. Since the basis blades are already represented in canonical order, the reversion can be computed by determining the grade of each blade (i.e. the length of the tuple) and multiplying the corresponding coefficient by the factor

$$(-1)^{\frac{k(k-1)}{2}}.$$

The following Python function illustrates the computational procedure. The function computes the reversion of a multivector by iterating through each term, calculating the grade, determining the appropriate factor, and producing the new multivector representation.

```
def reverse_multivector(mv):
    """
    Compute the reversion of a multivector.

    The multivector is assumed to be represented as a dictionary
        with keys
    as canonical basis blades (tuples of integers) and values as
        scalar coefficients.
    For each basis blade of grade k, the reversion multiplies the
        coefficient by the factor (-1)^(k*(k-1)/2).

    Parameters:
        mv (dict[tuple[int], float]): The input multivector.

    Returns:
        dict[tuple[int], float]: The reversion of the multivector.
    """
    reversed_mv = {}
    for blade, coeff in mv.items():
        grade = len(blade)
        factor = (-1) ** ((grade * (grade - 1)) // 2)
        reversed_mv[blade] = coeff * factor
    return reversed_mv
```

This function encapsulates the core concept of reversion. By calculating the grade of each blade from its tuple representation and applying the corresponding sign change, the function efficiently produces the reversed multivector. The use of integer division ensures that the exponent is computed correctly, in accordance with the algebraic definition.

Python Code Snippet

```python
# Comprehensive implementation of the reversion operation in
↪ Geometric Algebra,
# along with demonstration of its key properties and related
↪ algorithms.
#
# In this implementation, a multivector is represented as a
↪ dictionary,
# where the keys are basis blades represented as tuples of integers
↪ (in
# canonical sorted order) and values are the corresponding scalar
↪ coefficients.

def reverse_multivector(mv):
    """
    Compute the reversion (reversal) of a multivector.

    Every multivector can be expressed as a sum of homogeneous
    ↪ components.
    For a homogeneous blade of grade k, the reversion is given by:

        reverse(B) = (-1)^(k*(k-1)/2) * B

    This function applies the formula to each basis blade in the
    ↪ multivector.

    Parameters:
        mv (dict[tuple[int], float]): The input multivector.

    Returns:
        dict[tuple[int], float]: The multivector after applying
        ↪ reversion.
    """
    reversed_mv = {}
    for blade, coeff in mv.items():
        grade = len(blade)
        factor = (-1) ** ((grade * (grade - 1)) // 2)
        reversed_mv[blade] = coeff * factor
    return reversed_mv

def geometric_product(mv1, mv2):
    """
    Compute a simplistic version of the geometric product for two
    ↪ multivectors.

    Note:
        This is a naive implementation that concatenates the basis
        ↪ blades and
        sorts them to obtain a canonical form. In a full
        ↪ implementation, one
```

```
            would need to track sign changes due to reordering (using
         ↪  the anti-commutativity
            of basis vectors) and handle cancellations.

        Parameters:
            mv1 (dict[tuple[int], float]): First multivector.
            mv2 (dict[tuple[int], float]): Second multivector.

        Returns:
            dict[tuple[int], float]: The resulting multivector from the
         ↪  geometric product.
        """
        result = {}
        for blade1, coeff1 in mv1.items():
            for blade2, coeff2 in mv2.items():
                # Concatenate the blades and sort to form the canonical
             ↪  blade.
                new_blade = tuple(sorted(blade1 + blade2))
                result[new_blade] = result.get(new_blade, 0) + coeff1 *
             ↪  coeff2
        return result

def is_involution(mv):
        """
        Verify the involution property of the reversion operation.

        The reversion is an involution; that is, applying it twice
     ↪  should return
        the original multivector:

            reverse(reverse(mv)) == mv

        Parameters:
            mv (dict[tuple[int], float]): The input multivector.

        Returns:
            bool: True if the involution property holds, False
         ↪  otherwise.
        """
        return reverse_multivector(reverse_multivector(mv)) == mv

def print_multivector(mv):
        """
        Pretty-print the multivector in a readable format.

        Each term is printed in the form:
            coefficient * e(blade)
        where a scalar (grade 0) is printed without basis vector.

        Parameters:
            mv (dict[tuple[int], float]): The multivector to be printed.
        """
        terms = []
```

```python
    for blade, coeff in mv.items():
        if blade:
            term = f"{coeff}e{blade}"
        else:
            term = f"{coeff}"
        terms.append(term)
    print(" + ".join(terms))

# Example usage and demonstration of the reversion algorithm and
    properties
if __name__ == '__main__':
    # Define a multivector with components in various grades:
    #   Scalar part (grade 0), vectors (grade 1) and bivectors
        (grade 2).
    mv = {
        (): 3.0,            # Scalar component
        (1,): 2.0,          # Vector component
        (2,): -1.0,         # Vector component
        (1,2): 4.0,         # Bivector component
        (1,3): -2.0,        # Bivector component
        (2,3): 5.0          # Bivector component
    }

    print("Original multivector:")
    print_multivector(mv)

    # Compute the reversion of the multivector.
    mv_reversed = reverse_multivector(mv)
    print("\nReversed multivector:")
    print_multivector(mv_reversed)

    # Verify the involution property: reverse(reverse(mv)) should
        equal the original multivector.
    if is_involution(mv):
        print("\nInvolution property holds: reverse(reverse(mv)) ==
            mv")
    else:
        print("\nInvolution property does NOT hold.")

    # Demonstrate the anti-automorphism property of the reversion
        operation.
    # For any two multivectors A and B, reverse(A*B) should equal
        reverse(B)*reverse(A).
    A = {(1,): 1.0, (2,): 2.0}   # Example multivector A (vector
        components)
    B = {(2,): 3.0, (3,): -1.0}  # Example multivector B (vector
        components)

    AB = geometric_product(A, B)
    reversed_AB = reverse_multivector(AB)
    reversed_B = reverse_multivector(B)
    reversed_A = reverse_multivector(A)
    BA_reversed = geometric_product(reversed_B, reversed_A)
```

```python
print("\nGeometric product A * B:")
print_multivector(AB)

print("\nReversed geometric product reverse(A * B):")
print_multivector(reversed_AB)

print("\nGeometric product of reverse(B) and reverse(A):")
print_multivector(BA_reversed)

# Note:
# The geometric_product function here is a simplified
    representation.
# A full implementation would carefully handle the sign
    adjustments
# due to anti-commutativity when reordering basis vectors.
```

Chapter 10

Grade Involution: Signatures and Effects

Mathematical Formulation of Grade Involution

A multivector M in an n-dimensional geometric algebra admits a decomposition into homogeneous components

$$M = \sum_{k=0}^{n} \langle M \rangle_k,$$

where $\langle M \rangle_k$ denotes the grade-k component. The grade involution, denoted by a hat as in $\hat{M}$, is defined by applying a sign factor of $(-1)^k$ to each homogeneous part:

$$\hat{M} = \sum_{k=0}^{n} (-1)^k \langle M \rangle_k.$$

Under this operation, scalar components (grade 0) remain invariant, whereas odd-grade elements (such as vectors and trivectors) change sign and even-grade elements (such as bivectors and four-vectors) retain their sign. The involutive nature of this mapping is evident from the relation

$$\hat{\hat{M}} = M.$$

Algebraic Impact of Grade Involution

The grade involution operation induces a parity transformation that is central to the structure of geometric algebra. By multiplying each grade-k component by $(-1)^k$, the operation effectively distinguishes between elements of even and odd signature. This transformation has wide-ranging algebraic consequences. In particular, when considering the geometric product of multivectors, the grade involution distributes according to the relation

$$\widehat{AB} = \hat{A}\hat{B},$$

provided that the individual components are homogeneous. This property underpins many derivations within the algebra and facilitates the separation of even subalgebras, which often play a critical role in modeling rotations and reflections. The parity-dependent sign change realized by the grade involution also serves to reveal symmetry properties and simplify the analysis of composite multivector expressions, making it a valuable operation in both theoretical derivations and computational implementations.

Computational Implementation of Grade Involution

In computational contexts, multivectors are frequently represented as dictionaries where each key is a tuple corresponding to a canonical basis blade and each value is the associated scalar coefficient. The grade involution is then implemented by iterating over each basis blade, determining its grade (the length of the tuple), and multiplying the corresponding coefficient by $(-1)^k$. The following Python function encapsulates this procedure.

```
def grade_involution(mv):
    """
    Compute the grade involution of a multivector.

    The multivector is represented as a dictionary where each key is
      a tuple
    representing a basis blade and each value is the corresponding
      scalar coefficient.
    The grade involution operation multiplies the coefficient of
      each blade by (-1)^k,
    where k is the grade of the blade (i.e., the length of the
      tuple).
```

```
Parameters:
    mv (dict[tuple[int], float]): The input multivector.

Returns:
    dict[tuple[int], float]: The multivector after applying the
    ↪ grade involution.
"""
result = {}
for blade, coeff in mv.items():
    grade = len(blade)
    result[blade] = coeff * ((-1) ** grade)
return result
```

This function reflects the precise algebraic definition by processing each component of the multivector individually. The use of a dictionary to represent the multivector enables efficient access and modification of each term, ensuring that the grade involution is performed in a manner that is both computationally transparent and faithful to the underlying mathematical structure.

Python Code Snippet

```
"""
This module implements key algorithms and formulas discussed in the
↪ chapter on Grade Involution.
Multivectors are represented as dictionaries where each key is a
↪ tuple representing a basis blade
(e.g., () for scalars, (1,) for vector e1, (1,2) for bivector e12,
↪ etc.) and each value is the associated
scalar coefficient.

The implemented routines include:
  - grade_involution: Applying the sign (-1)^grade to each
  ↪ homogeneous component.
  - gp_basis: Computing the geometric product of two basis blades.
  - geometric_product: Distributing the geometric product over
  ↪ multivectors.
  - Utility routines and test cases to verify properties such as:
      * Involution: grade_involution(grade_involution(M)) == M
      * Distribution over the geometric product for homogeneous
         ↪ multivectors:
           grade_involution(A * B) == grade_involution(A) *
           ↪ grade_involution(B)
"""

def grade_involution(mv):
    """
```

```
    Compute the grade involution of a multivector.

    The multivector is represented as a dictionary where each key
    ↪  represents a basis blade
    (a tuple of integers) and each value is its coefficient. The
    ↪  grade involution multiplies
    each component by (-1)^grade, where grade is the length of the
    ↪  tuple.

    Parameters:
        mv (dict[tuple, float]): Input multivector.

    Returns:
        dict[tuple, float]: Multivector after applying grade
        ↪  involution.
    """
    result = {}
    for blade, coeff in mv.items():
        grade = len(blade)
        result[blade] = coeff * ((-1) ** grade)
    return result

def gp_basis(blade1, blade2):
    """
    Compute the geometric product of two basis blades.

    The basis blades (represented as tuples) are multiplied by
    ↪  concatenating the tuples,
    then reordering the result to a sorted tuple. The process
    ↪  considers the anti-commuting
    nature of the basis vectors, counting the number of
    ↪  transpositions, and removes duplicate
    indices (using e_i^2 = 1 for a Euclidean signature).

    Parameters:
        blade1 (tuple): First basis blade (assumed to be in sorted
        ↪  order).
        blade2 (tuple): Second basis blade (assumed to be in sorted
        ↪  order).

    Returns:
        (int, tuple): A tuple containing the sign factor and the
        ↪  resulting basis blade.
    """
    result_blade = list(blade1)
    sign = 1
    for b in blade2:
        # Determine the correct insertion index for 'b' in the
        ↪  sorted order.
        idx = 0
        while idx < len(result_blade) and result_blade[idx] < b:
            idx += 1
        if idx < len(result_blade) and result_blade[idx] == b:
```

```python
                    # If 'b' is already present, remove it (since e_i*e_i =
                    ↪  1).
                    result_blade.pop(idx)
                    # No sign change on removal.
                else:
                    # Number of transpositions is the number of elements
                    ↪  after the insertion point.
                    swaps = len(result_blade) - idx
                    sign *= (-1) ** swaps
                    result_blade.insert(idx, b)
        return sign, tuple(result_blade)

def geometric_product(mv1, mv2):
    """
    Compute the geometric product of two multivectors.

    The geometric product is computed by distributing over each term
    ↪  of the multivectors.
    For each pair of basis blades (from mv1 and mv2), the product is
    ↪  determined using gp_basis.

    Parameters:
        mv1 (dict[tuple, float]): First multivector.
        mv2 (dict[tuple, float]): Second multivector.

    Returns:
        dict[tuple, float]: The resulting multivector from the
        ↪  geometric product.
    """
    result = {}
    for blade1, coeff1 in mv1.items():
        for blade2, coeff2 in mv2.items():
            s, blade_res = gp_basis(blade1, blade2)
            combined_coeff = coeff1 * coeff2 * s
            if blade_res in result:
                result[blade_res] += combined_coeff
            else:
                result[blade_res] = combined_coeff
    return result

def multivector_equal(mv1, mv2, tol=1e-10):
    """
    Check if two multivectors are equal within a specified
    ↪  tolerance.

    This utility function cleans near-zero coefficients and compares
    ↪  the two multivectors.

    Parameters:
        mv1, mv2 (dict[tuple, float]): Multivectors to compare.
        tol (float): Tolerance level for considering a coefficient
        ↪  as zero.
```

```
    Returns:
        bool: True if the multivectors are equal within the given
        ↪   tolerance, False otherwise.
    """
    def clean(mv):
        return {blade: coeff for blade, coeff in mv.items() if
        ↪   abs(coeff) > tol}

    return clean(mv1) == clean(mv2)

# ------------------------ Testing the Implementations
↪   ------------------------

# Define a sample multivector M with mixed grades:
# M = 3 + 2e1 - 4e12 + 7e123
M = {
    (): 3,          # Scalar part (grade 0)
    (1,): 2,        # Vector part (grade 1)
    (1,2): -4,      # Bivector part (grade 2)
    (1,2,3): 7      # Trivector part (grade 3)
}

# Apply grade involution to M
M_involuted = grade_involution(M)
# Apply grade involution again to test involutive property: ^(^M) =
↪   M
M_double_involuted = grade_involution(M_involuted)

print("Original M:", M)
print("Grade involution of M:", M_involuted)
print("Double involution of M (should equal original M):",
↪   M_double_involuted)
print("Double involution matches original:", multivector_equal(M,
↪   M_double_involuted))

# Testing the distribution property of grade involution over the
↪   geometric product:
# For homogeneous multivectors A and B, we expect:
# grade_involution(A*B) == grade_involution(A) * grade_involution(B)

# Define two homogeneous vector multivectors:
A = {(1,): 1}   # Represents vector e1 (grade 1)
B = {(2,): 1}   # Represents vector e2 (grade 1)

# Compute their geometric product (A * B should yield bivector e12)
AB = geometric_product(A, B)

# Apply grade involution on the geometric product of A and B
hat_AB = grade_involution(AB)
# Compute the product of the grade involutions of A and B separately
hat_A = grade_involution(A)
hat_B = grade_involution(B)
product_hatA_hatB = geometric_product(hat_A, hat_B)
```

```
print("\nA (e1):", A)
print("B (e2):", B)
print("Geometric product A*B (expected bivector e12):", AB)
print("Grade involution of A*B:", hat_AB)
print("Geometric product of grade involutions (hat(A)*hat(B)):",
↪   product_hatA_hatB)
print("Distribution property holds:", multivector_equal(hat_AB,
↪   product_hatA_hatB))
```

Chapter 11

Clifford Conjugation: Techniques and Applications

Mathematical Foundations of Clifford Conjugation

In a geometric algebra defined over an n-dimensional vector space, a multivector M admits a unique decomposition into its homogeneous components,

$$M = \sum_{k=0}^{n} \langle M \rangle_k,$$

where $\langle M \rangle_k$ denotes the grade-k part of M. Clifford conjugation is defined as the composition of two involutive operations: the grade involution and the reversion. Denoting the grade involution by $\hat{M}$ and the reversion by $\widetilde{M}$, Clifford conjugation is expressed as

$$M^* = \widetilde{\hat{M}}.$$

The algebraic impact of these operations on each grade-k component of M is encapsulated by a sign factor in the form

$$(-1)^{\frac{k(k+1)}{2}},$$

so that the conjugated multivector takes the explicit form

$$M^* = \sum_{k=0}^{n} (-1)^{\frac{k(k+1)}{2}} \langle M \rangle_k.$$

This graded sign assignment emerges naturally from the combination of the reversal's permutation of basis vectors and the grade involution's parity transformation. The resultant operation is pivotal in addressing symmetry properties and computational simplifications within the algebra.

Algebraic Simplification and Symmetry Analysis

The structured modification of signs provided by Clifford conjugation yields significant algebraic simplifications. In particular, for any two homogeneous multivectors A and B, Clifford conjugation satisfies the relation

$$(AB)^* = B^* A^*.$$

This reversal of the order, accompanied by the systematic sign change across different grades, facilitates the derivation of invariant forms and the detection of underlying symmetries in multivector products. Such properties are of paramount importance in the study of physical systems where reflection and rotation symmetries are central, and in computational frameworks that benefit from a reduction in the complexity of symbolic and numerical expressions.

Clifford conjugation often serves as a tool to simplify the solution of algebraic equations in geometric algebra by canceling out asymmetric contributions or grouping terms in a manner that preserves physical invariance. The clear separation of contributions from even and odd-grade elements further assists in the identification of conserved quantities and invariant subalgebras within the overall structure.

Computational Implementation of Clifford Conjugation

The computational treatment of multivectors frequently employs a dictionary-based representation, where each key is a tuple encoding a basis blade and the corresponding value represents its scalar coefficient. An efficient implementation of Clifford conjugation involves iterating through this dictionary and applying the sign factor $(-1)^{\frac{k(k+1)}{2}}$ to each term, where k is determined by the length of the tuple corresponding to a basis blade.

The function below encapsulates the procedure for Clifford conjugation on a multivector. The accompanying code snippet is provided in a minted environment for clarity.

```python
def clifford_conjugation(mv):
    """
    Compute the Clifford conjugation of a multivector.
    
    The multivector is represented as a dictionary where each key is
      a tuple
    corresponding to a canonical basis blade, and the associated
      value is the coefficient.
    Clifford conjugation applies a sign of (-1)^(k*(k+1)/2) to the
      grade-k component,
    where k is the length of the tuple representing the blade.
    
    Parameters:
        mv (dict[tuple, float]): Input multivector.
    
    Returns:
        dict[tuple, float]: Multivector after applying Clifford
          conjugation.
    """
    result = {}
    for blade, coeff in mv.items():
        k = len(blade)
        sign = (-1) ** (k * (k + 1) // 2)
        result[blade] = coeff * sign
    return result
```

Python Code Snippet

```python
class Multivector:
    def __init__(self, components=None):
        """
        Initialize a multivector.
        
        The multivector is stored as a dictionary where:
          - Each key is a tuple representing a basis blade (e.g., ()
            for scalar, (1,) for e1, (1,2) for e1^e2, etc.)
          - Each value is the numerical coefficient.
        """
        if components is None:
```

```python
        self.components = {}
    else:
        # Ensure a copy is stored to avoid accidental mutation.
        self.components = components.copy()

def __add__(self, other):
    """
    Add two multivectors.
    """
    result = self.components.copy()
    for blade, coeff in other.components.items():
        result[blade] = result.get(blade, 0.0) + coeff
    # Optionally remove near-zero coefficients
    result = {blade: c for blade, c in result.items() if abs(c)
      ↪ > 1e-12}
    return Multivector(result)

def __neg__(self):
    """
    Negate a multivector.
    """
    return Multivector({blade: -coeff for blade, coeff in
      ↪ self.components.items()})

def __sub__(self, other):
    """
    Subtract two multivectors.
    """
    return self + (-other)

def __repr__(self):
    """
    Return a string representation of the multivector.
    """
    if not self.components:
        return "0"
    terms = []
    for blade, coeff in sorted(self.components.items(),
      ↪ key=lambda x: (len(x[0]), x[0])):
        if blade == ():
            terms.append(f"{coeff}")
        else:
            terms.append(f"{coeff}*e{blade}")
    return " + ".join(terms)

def grade(self, k):
    """
    Extract the grade-k part of the multivector.
    """
    return Multivector({blade: coeff for blade, coeff in
      ↪ self.components.items() if len(blade) == k})

def grade_involution(self):
```

```
    """
    Compute the grade involution of the multivector.

    This operation multiplies each grade-k component by (-1)^k.
    """
    new_components = {}
    for blade, coeff in self.components.items():
        k = len(blade)
        new_components[blade] = coeff * ((-1) ** k)
    return Multivector(new_components)

def reversion(self):
    """
    Compute the reversion of the multivector.

    Reversion multiplies each grade-k component by
    ↪  (-1)^(k*(k-1)/2).
    """
    new_components = {}
    for blade, coeff in self.components.items():
        k = len(blade)
        factor = (-1) ** (k * (k - 1) // 2)
        new_components[blade] = coeff * factor
    return Multivector(new_components)

def clifford_conjugate(self):
    """
    Compute the Clifford conjugation of the multivector.

    Clifford conjugation is defined as the composition of the
    ↪  grade involution
    and the reversion. The combined effect is to multiply a
    ↪  grade-k component by (-1)^(k + k*(k-1)/2)
    which simplifies to (-1)^(k*(k+1)/2).
    """
    # Alternatively, one could compute sign = (-1)^(k*(k+1)/2)
    ↪  directly,
    # but here we combine the two operations for clarity.
    return self.grade_involution().reversion()

def geometric_product(self, other):
    """
    Compute the geometric product of two multivectors.

    This implementation assumes a Euclidean metric where
     - Basis vectors square to 1 (e_i^2 = 1).
     - Distinct basis vectors anticommute (e_i * e_j = -e_j *
    ↪  e_i for i != j).

    The product is computed blade-by-blade using the helper
    ↪  function 'blade_product'.
    """
    result = {}
```

```python
        for blade1, coeff1 in self.components.items():
            for blade2, coeff2 in other.components.items():
                new_blade, sign = self.blade_product(blade1, blade2)
                key = new_blade
                result[key] = result.get(key, 0.0) + coeff1 * coeff2
                ↪   * sign
        result = {blade: c for blade, c in result.items() if abs(c)
        ↪   > 1e-12}
        return Multivector(result)

    @staticmethod
    def blade_product(blade1, blade2):
        """
        Compute the geometric product of two basis blades.

        Parameters:
            blade1 (tuple): A tuple representing the first basis
            ↪   blade.
            blade2 (tuple): A tuple representing the second basis
            ↪   blade.

        Returns:
            A tuple (new_blade, sign) where:
                - new_blade is the resulting basis blade after
                ↪   multiplication.
                - sign is the accumulated sign due to anticommutation
                ↪   and cancellation of duplicate basis vectors.

        The algorithm:
            1. Concatenate the blades.
            2. Sort the concatenated list and count the number of
            ↪   swaps (each swap contributes a factor of -1).
            3. Remove duplicate basis vectors (since e_i^2 = 1,
            ↪   duplicates cancel out).
        """
        # Combine the two blades
        blade_list = list(blade1) + list(blade2)
        sign = 1
        # Bubble sort to count the number of swaps needed to sort
        ↪   the basis indices
        for i in range(len(blade_list)):
            for j in range(0, len(blade_list) - i - 1):
                if blade_list[j] > blade_list[j + 1]:
                    blade_list[j], blade_list[j + 1] = blade_list[j
                    ↪   + 1], blade_list[j]
                    sign *= -1
        # Remove pairs of equal basis vectors (since e_i^2 = 1)
        i = 0
        reduced_blade = []
        while i < len(blade_list):
            if i + 1 < len(blade_list) and blade_list[i] ==
            ↪   blade_list[i + 1]:
                # The pair cancels to 1
```

89

```python
                i += 2
            else:
                reduced_blade.append(blade_list[i])
                i += 1
        return tuple(reduced_blade), sign

# Example usage and verification of properties
if __name__ == "__main__":
    # Define some basic multivectors
    # Scalar: 2.0
    scalar = Multivector({(): 2.0})

    # Vectors: e1 and e2 with coefficients 1.0 and 3.0 respectively
    vector1 = Multivector({(1,): 1.0})
    vector2 = Multivector({(2,): 3.0})

    # Compute the bivector: e1^e2 as the geometric product of
    #    vector1 and vector2
    bivector = vector1.geometric_product(vector2)

    # Construct an arbitrary multivector M = scalar + vector1 +
    #    vector2 + bivector
    M = scalar + vector1 + vector2 + bivector
    print("Multivector M:")
    print(M)

    # Compute the Clifford conjugation of M
    M_cc = M.clifford_conjugate()
    print("\nClifford Conjugation of M:")
    print(M_cc)

    # Demonstrate the property: (AB)* = B* A*
    # Define multivectors A and B
    A = vector1 + vector2          # A simple vector multivector
    #    (homogeneous, grade 1)
    B = scalar + bivector          # A combination of scalar and
    #    bivector (grades 0 and 2)

    # Compute the geometric product AB
    AB = A.geometric_product(B)

    # Compute Clifford conjugation on the product and individually
    #    on A and B
    left_side = AB.clifford_conjugate()
    right_side =
    ↪   B.clifford_conjugate().geometric_product(A.clifford_conjugate())

    print("\nVerifying (AB)* = B* A*:")
    print("Left side (AB)*:")
    print(left_side)
    print("\nRight side B* A*:")
    print(right_side)
```

Chapter 12

Norms in Geometric Algebra: Measuring Magnitudes

Mathematical Framework of Norms in Geometric Algebra

A multivector in geometric algebra embodies a rich structure that permits several definitions of a norm. One commonly adopted definition measures the magnitude of a multivector by considering the scalar part of its product with its Clifford conjugate. This is expressed mathematically by

$$\|M\| = \sqrt{\left|\langle MM^*\rangle_0\right|},$$

where M^* denotes the Clifford conjugate of M and $\langle \cdot \rangle_0$ represents the extraction of the scalar (grade zero) component. In the special case when M is a scalar, the norm reduces to its absolute value, while in more elaborate configurations the norm provides a measure that respects the contributions of each grade as dictated by the underlying metric structure. The systematic sign adjustments introduced during the Clifford conjugation enable this norm to inherit properties analogous to those of standard Euclidean norms while extending naturally to blades and multivectors of higher grade.

Physical Significance of Norms

In physical applications, the norm of an element in geometric algebra plays a pivotal role in quantitatively describing the magnitude of physical quantities. For instance, in the analysis of rotations and reflections, the normalization of rotors and spinors is achieved through the careful determination of their associated norms. Norms also appear in formulations of conservation laws: by evaluating the invariance of the scalar part of a multivector product under specific transformation groups, one can deduce conserved quantities and symmetry invariants in physical systems.

The capability to measure lengths, areas, and volumes within a unified algebraic framework offers computational and conceptual advantages. These aspects are particularly evident in simulations of dynamical systems, where the preservation of magnitude under time evolution is crucial, and in quantum mechanics, where the normalization of wave functions is intimately connected to probabilistic interpretations.

Computational Implementation of Norms in Geometric Algebra

The efficient calculation of norms in computational applications frequently involves representing multivectors as dictionaries, with keys corresponding to basis blades and values representing the associated coefficients. Such a representation enables the systematic execution of algebraic operations by mapping each basis blade to its numerical weight.

A practical approach computes the norm by first obtaining the Clifford conjugate of the multivector, then calculating the geometric product of the multivector with this conjugate, and finally extracting the scalar part. The norm is then defined as the square root of the absolute value of this scalar. The function below encapsulates this procedure for a multivector stored in dictionary format:

```
def compute_ga_norm(mv):
    """
    Calculate the norm of a multivector in geometric algebra.
    The norm is defined as the square root of the absolute value
    of the scalar part of the product of the multivector with its
    Clifford conjugate, expressed as:
```

```
                 ||M|| = sqrt(|<M * M*>_0|)

    Parameters:
        mv (dict): A dictionary representing the multivector, where
                   each key is a tuple corresponding to a basis
                   ↪ blade and
                   the associated value is its coefficient.

    Returns:
        float: The computed norm of the multivector.
    """
    mv_cc = clifford_conjugation(mv)      # Compute the Clifford
    ↪ conjugate of mv.
    product = geometric_product(mv, mv_cc)   # Compute the geometric
    ↪ product.
    scalar_part = product.get((), 0.0)       # Retrieve the scalar
    ↪ part stored under the empty tuple key.
    return abs(scalar_part) ** 0.5
```

This implementation highlights the utilization of structured data representations for multivectors and demonstrates how algebraic operations are leveraged to compute measurable magnitudes. By focusing on the isolation of the zero-grade component, the method remains robust even in the presence of a diverse set of basis elements, ensuring that the computed norm retains its physical and computational significance.

Python Code Snippet

```
# Comprehensive Python implementation for computing norms in
↪ Geometric Algebra (GA)
# using a dictionary-based representation of multivectors.
#
# In this implementation:
# - Multivectors are represented as dictionaries where keys are
↪ tuples representing
#   basis blades (e.g., () for the scalar part, (1,) for e1, (1,2)
↪ for e12, etc.)
# - The geometric product of two multivectors is computed by
↪ multiplying their basis
#   blades with proper sign adjustments.
# - The Clifford conjugation applies a sign factor (-1)^(r(r+1)/2)
↪ to each blade
#   of grade r.
# - The norm of a multivector M is given by ||M|| = sqrt(|<M *
↪ M*>_0|), where M* is
#   the Clifford conjugate and the scalar part is extracted using
↪ the empty tuple key ().
```

```python
def sort_with_sign(lst):
    """
    Sorts a list of integers and returns the sorted list along with
    the permutation sign (1 for even number of swaps, -1 for odd
    ↪ swaps).

    Parameters:
        lst (list): List of integers representing basis indices.

    Returns:
        tuple: A tuple (sorted_list, sign) where sorted_list is the
        ↪ sorted version of lst
            and sign is 1 or -1 depending on the parity of the
            ↪ permutation.
    """
    arr = lst.copy()
    n = len(arr)
    sign = 1
    # Use a simple bubble sort to count swaps
    for i in range(n):
        for j in range(0, n - i - 1):
            if arr[j] > arr[j + 1]:
                arr[j], arr[j + 1] = arr[j + 1], arr[j]
                sign = -sign
    return arr, sign

def cancel_duplicates(sorted_list):
    """
    Cancels out duplicate adjacent basis indices from a sorted list.
    In an orthonormal Euclidean metric, e_i * e_i = 1 so pairs
    ↪ cancel.

    Parameters:
        sorted_list (list): A sorted list of basis indices.

    Returns:
        tuple: A tuple representing the resulting basis blade after
        ↪ cancellation.
    """
    result = []
    i = 0
    n = len(sorted_list)
    while i < n:
        # If the next element is identical, cancel the pair
        if i + 1 < n and sorted_list[i] == sorted_list[i + 1]:
            i += 2
        else:
            result.append(sorted_list[i])
            i += 1
    return tuple(result)

def geometric_product_blades(blade1, blade2):
```

```
    """
    Computes the geometric product of two basis blades.

    The procedure is as follows:
        1. Concatenate the lists representing the two blades.
        2. Sort the combined list while counting the number of
        ↪  interchanges,
           which yields an overall sign factor.
        3. Cancel out pairs of identical basis indices using the
        ↪  identity e_i * e_i = 1.

    Parameters:
        blade1 (tuple): A tuple representing the first basis blade.
        blade2 (tuple): A tuple representing the second basis blade.

    Returns:
        tuple: A tuple (result_blade, sign) where result_blade is
        ↪  the resulting basis
              blade as a tuple and sign is the accumulated sign
              ↪  factor from the permutation.
    """
    combined = list(blade1) + list(blade2)
    sorted_comb, sign = sort_with_sign(combined)
    result_blade = cancel_duplicates(sorted_comb)
    return result_blade, sign

def geometric_product(mv1, mv2):
    """
    Computes the geometric product of two multivectors.

    Each multivector is represented as a dictionary with keys as
    ↪  tuples (basis blades)
    and values as the corresponding coefficients.

    The product is computed by multiplying every term in the first
    ↪  multivector with
    every term in the second multivector. The product of the
    ↪  individual basis blades is
    computed using the geometric_product_blades function.

    Parameters:
        mv1 (dict): The first multivector.
        mv2 (dict): The second multivector.

    Returns:
        dict: A new multivector (dictionary) representing the
        ↪  geometric product.
    """
    result = {}
    for blade1, coeff1 in mv1.items():
        for blade2, coeff2 in mv2.items():
            new_blade, sign = geometric_product_blades(blade1,
            ↪  blade2)
```

```python
            new_coeff = coeff1 * coeff2 * sign
            if new_blade in result:
                result[new_blade] += new_coeff
            else:
                result[new_blade] = new_coeff
    return result

def clifford_conjugation(mv):
    """
    Computes the Clifford conjugate of a multivector.

    For each basis blade of grade r (i.e., r = len(blade)), the
    ↪ Clifford conjugate
    is defined as:
        e_A* = (-1)^(r(r+1)/2) e_A
    This operation applies a systematic sign change to the
    ↪ components of the multivector.

    Parameters:
        mv (dict): The original multivector.

    Returns:
        dict: A new multivector representing the Clifford conjugate
        ↪ of mv.
    """
    result = {}
    for blade, coeff in mv.items():
        grade = len(blade)
        sign = (-1) ** (grade * (grade + 1) // 2)
        result[blade] = coeff * sign
    return result

def compute_ga_norm(mv):
    """
    Computes the norm of a multivector in geometric algebra.

    The norm is given by:
        ||M|| = sqrt(|<M * M*>_0|)
    where M* is the Clifford conjugate of M and < . >_0 extracts the
    ↪ scalar (grade 0)
    portion of the product.

    Parameters:
        mv (dict): The multivector.

    Returns:
        float: The computed norm.
    """
    # Step 1: Compute the Clifford conjugate of mv
    mv_cc = clifford_conjugation(mv)
    # Step 2: Compute the geometric product of mv and its Clifford
    ↪ conjugate
    product = geometric_product(mv, mv_cc)
```

```python
    # Step 3: Extract the scalar part, stored under the key ().
    scalar_part = product.get((), 0.0)
    # Step 4: Return the square root of the absolute value of the
    ↪  scalar part.
    return abs(scalar_part) ** 0.5

# Example usage
if __name__ == "__main__":
    # Define a sample multivector in 3D geometric algebra:
    # Let M = 3 + 2e1 - e2 + 0.5e12 + e123.
    # Representation:
    #   ()        : scalar component (3)
    #   (1,)      : e1 component (2)
    #   (2,)      : e2 component (-1)
    #   (1,2)     : e12 component (0.5)
    #   (1,2,3)   : e123 component (1)
    mv = {
        (): 3.0,
        (1,): 2.0,
        (2,): -1.0,
        (1,2): 0.5,
        (1,2,3): 1.0
    }

    norm_value = compute_ga_norm(mv)
    print("The norm of the multivector M is:", norm_value)
```

Chapter 13

Inversion of Multivectors: Computational Methods

Mathematical Framework for Multivector Inversion

Let a general multivector M in a Clifford algebra be expressed as a sum of homogeneous grade elements. The inverse M^{-1}, when it exists, is defined by the relation

$$MM^{-1} = M^{-1}M = 1,$$

where 1 denotes the scalar identity of the algebra. A standard method for constructing the inverse utilizes the Clifford conjugate, denoted by M^*, and the scalar part obtained via the grade projection operator $\langle \cdot \rangle_0$. In many Clifford algebra settings the inverse is given by the formula

$$M^{-1} = \frac{M^*}{\langle M M^* \rangle_0},$$

provided that $\langle M M^* \rangle_0 \neq 0$. The scalar $\langle M M^* \rangle_0$ acts as a normalization factor, ensuring that the product of M and its inverse yields the identity element. The effectiveness of this approach rests on the properties of the geometric product, including associativity and distributivity, along with the involutive nature of the Clifford conjugation.

Algorithmic Considerations and Inversion Criteria

The computation of the inverse of a multivector necessitates several algorithmic steps. Initially, the Clifford conjugate M^* is calculated. Following this, the geometric product $M M^*$ is computed. The extraction of the scalar component $\langle M M^* \rangle_0$ is an essential step; it determines both the invertibility and the appropriate scaling factor needed for inversion. In practical implementations, multivectors are frequently represented in a sparse data structure (for instance, using dictionaries where keys correspond to basis blades), and arithmetic operations are implemented to track the sign changes and cancellations that occur during the geometric product.

A key aspect of the algorithm is to ensure floating-point stability when evaluating $\langle M M^* \rangle_0$. If this scalar is zero (or nearly zero within an accepted tolerance), then the multivector does not possess a multiplicative inverse within the algebra. Otherwise, the division by this scalar yields an inverse that satisfies the condition $M M^{-1} = 1$.

Computational Implementation of Multivector Inversion

An implementation of the inversion procedure can be encapsulated in a compact Python function. This function assumes that auxiliary operations such as the Clifford conjugation and the geometric product—responsible for handling the multivector arithmetic—are available. The following code snippet presents a function that computes the inverse of a multivector represented as a dictionary. The keys correspond to basis blades (expressed as tuples) and the values denote the coefficients.

```
def invert_multivector(mv):
    """
```

```
Compute the inverse of a multivector in geometric algebra.

The inverse M_inv is defined by the relation
    M * M_inv = M_inv * M = 1,
where 1 is the scalar identity. The algorithm uses the Clifford
↪ conjugate
(M_star) and the scalar part of the geometric product, as given
↪ by:
    M_inv = M_star / <M * M_star>_0,
provided that <M * M_star>_0 is non-zero.

Parameters:
    mv (dict): A dictionary representation of the multivector,
               where each key is a tuple corresponding to a
               ↪ basis blade,
               and the associated value is its coefficient.

Returns:
    dict or None: The inverse multivector if it exists;
    ↪ otherwise, None.
"""
mv_star = clifford_conjugation(mv)   # Compute the Clifford
↪ conjugate of mv.
product = geometric_product(mv, mv_star)   # Compute the
↪ geometric product M * M_star.
scalar_part = product.get((), 0.0)   # Extract the scalar part
↪ from the result.

# Check for invertibility based on the non-vanishing scalar
↪ component.
if abs(scalar_part) < 1e-12:
    return None   # The multivector is not invertible.

# Construct the inverse by scaling the conjugate.
inverse = {}
for blade, coeff in mv_star.items():
    inverse[blade] = coeff / scalar_part
return inverse
```

Python Code Snippet

```
# Comprehensive Python implementation for the inversion of a
↪ multivector in a
# Clifford algebra. In this implementation, multivectors are
↪ represented as
# dictionaries where keys are tuples representing basis blades
↪ (e.g., () for the
# scalar, (1,) for e1, (1,2) for e1 e2, etc.) and values are their
↪ coefficients.
#
```

```
# The inversion is computed using the formula:
#       M^{-1} = M^{*} / <M * M^{*}>_0,
# where M^{*} is the Clifford conjugate of the multivector M and
# <M * M^{*}>_0 is the scalar part of the geometric product M *
↪  M^{*}.
#
# Assumptions:
#  - An orthonormal basis is used (e_i^2 = 1 for each basis vector).
#  - The geometric product is implemented via the helper function
↪  'multiply_blades',
#    which computes the product of two basis blades taking into
↪  account the
#    anti-commutation rules and sign changes.
#
# The code below includes functions for:
#  1. multiply_blades     : Computes the product of two basis
↪  blades.
#  2. geometric_product   : Computes the geometric product of two
↪  multivectors.
#  3. clifford_conjugation: Computes the Clifford conjugate
↪  (reversal with a
#                           sign factor) of a multivector.
#  4. invert_multivector  : Computes the inverse of a multivector,
↪  if it exists.
#  5. An example usage demonstrating the inversion and
↪  verification.

def multiply_blades(b1, b2):
    """
    Multiply two basis blades in an orthonormal Clifford algebra.

    Parameters:
        b1 (tuple): A tuple representing the first basis blade.
        b2 (tuple): A tuple representing the second basis blade.

    Returns:
        (tuple, int): A tuple containing the resulting basis blade
↪  (as a sorted tuple)
                      and the sign obtained from the
↪  anti-commutation relations.
    """
    blade_list = list(b1)  # Start with the first blade as a mutable
↪  list.
    sign = 1
    # Process each basis element in the second blade.
    for v in b2:
        if v in blade_list:
            # If v is already present, remove it (since e_i * e_i =
↪  1) without a sign change.
            blade_list.remove(v)
        else:
            # Determine the correct insertion position to keep the
↪  list sorted.
```

```python
            pos = 0
            while pos < len(blade_list) and blade_list[pos] < v:
                pos += 1
            # The number of swaps needed is the number of elements
            ↪ after the insertion point.
            swaps = len(blade_list) - pos
            sign *= (-1) ** swaps
            blade_list.insert(pos, v)
    return (tuple(blade_list), sign)

def geometric_product(mv1, mv2):
    """
    Compute the geometric product of two multivectors.

    Parameters:
        mv1 (dict): The first multivector, with keys as basis blades
        ↪ (tuples)
                and values as coefficients (floats).
        mv2 (dict): The second multivector, in the same dictionary
        ↪ format.

    Returns:
        dict: The resulting multivector from the geometric product.
    """
    result = {}
    # Iterate through all pairs of basis blades from mv1 and mv2.
    for blade1, coeff1 in mv1.items():
        for blade2, coeff2 in mv2.items():
            prod_blade, sign = multiply_blades(blade1, blade2)
            new_coeff = coeff1 * coeff2 * sign
            if prod_blade in result:
                result[prod_blade] += new_coeff
            else:
                result[prod_blade] = new_coeff
    # Filter out any near-zero terms.
    result = {blade: coeff for blade, coeff in result.items() if
    ↪ abs(coeff) > 1e-12}
    return result

def clifford_conjugation(mv):
    """
    Compute the Clifford conjugate of a multivector.

    The Clifford conjugate of a basis blade is obtained by reversing
    ↪ the order
    of the basis vectors and multiplying by a sign factor given by
        (-1)^(grade*(grade-1)/2),
    where 'grade' is the number of basis vectors in the blade.

    Parameters:
        mv (dict): A multivector represented as a dictionary.

    Returns:
```

            dict: The Clifford conjugate of the input multivector.
        """
        conjugated = {}
        for blade, coeff in mv.items():
            grade = len(blade)
            sign = (-1) ** (grade * (grade - 1) // 2)
            new_blade = tuple(reversed(blade))
            conjugated[new_blade] = coeff * sign
        return conjugated

def invert_multivector(mv):
    """
    Compute the inverse of a multivector in geometric algebra.

    The inverse M_inv is defined by:
        M * M_inv = M_inv * M = 1,
    where 1 is the scalar identity. The algorithm uses the Clifford
    ↪ conjugate,
    M_star, and the scalar part of the geometric product:
        M_inv = M_star / <M * M_star>_0,
    provided that <M * M_star>_0 is non-zero.

    Parameters:
        mv (dict): A dictionary representation of the multivector.

    Returns:
        dict or None: The inverse multivector if it exists;
        ↪ otherwise, None.
    """
    mv_star = clifford_conjugation(mv)   # Compute the Clifford
    ↪ conjugate of mv.
    product = geometric_product(mv, mv_star)   # Compute the product
    ↪ M * M_star.
    scalar_part = product.get((), 0.0)   # Extract the scalar
    ↪ (grade-0) part.

    # Check if the multivector is invertible.
    if abs(scalar_part) < 1e-12:
        return None   # Multivector is not invertible as the scalar
        ↪ part is zero.

    # Scale the conjugate by the scalar part to obtain the inverse.
    inverse = {}
    for blade, coeff in mv_star.items():
        inverse[blade] = coeff / scalar_part
    return inverse

# Example usage demonstrating the inversion of a multivector:
if __name__ == "__main__":
    # Define a sample multivector M = 2 + 3e1 + 4e2 + 5e1e2
    # Representation:
    #   ()   -> scalar part (2)
    #   (1,) -> e1 (3)

103

```python
#   (2,)  -> e2 (4)
#   (1,2) -> e1e2 (5)
M = {
    (): 2.0,
    (1,): 3.0,
    (2,): 4.0,
    (1,2): 5.0
}

# Compute the inverse of M.
M_inv = invert_multivector(M)

if M_inv is not None:
    print("The inverse of the multivector M is:")
    # Display the inverse multivector sorted by grade and basis
    #    blade.
    for blade, coeff in sorted(M_inv.items(), key=lambda x:
    →   (len(x[0]), x[0])):
        if blade == ():
            print(f"{coeff}")
        else:
            blade_str = "e" + "".join(str(i) for i in blade)
            print(f"{coeff} {blade_str}")
    # Verification: Compute M * M_inv to check if it equals the
    →   scalar identity.
    identity = geometric_product(M, M_inv)
    print("\nVerification (M * M_inv):")
    for blade, coeff in sorted(identity.items(), key=lambda x:
    →   (len(x[0]), x[0])):
        if blade == ():
            print(f"Scalar: {coeff}")
        else:
            blade_str = "e" + "".join(str(i) for i in blade)
            print(f"{blade_str}: {coeff}")
else:
    print("The multivector M is not invertible.")
```

Chapter 14

Exponential Functions in GA: Theory and Computation

The Theoretical Foundation

The exponential of a multivector M in a Clifford algebra is defined by the power series

$$\exp(M) = \sum_{n=0}^{\infty} \frac{M^n}{n!},$$

where M^0 is understood to be the scalar identity and M^n denotes the n-fold geometric product of M. The formulation of the exponential function relies on the associativity and distributivity of the geometric product. When M is decomposed into its homogeneous grade components, the convergence of the series is governed by the norm of M, and in many cases, the series can be separated into even and odd parts. For instance, when M is a bivector that satisfies $M^2 = -\theta^2$ for some scalar θ, the series can be reorganized into a trigonometric form:

$$\exp(M) = \cos(\theta) + \frac{M}{\theta} \sin(\theta).$$

This result is of particular significance in the context of rotations and Lorentz transformations in physics, wherein such closed forms of the exponential play a central role.

Decomposition and Closed-Form Expressions

The multivector M can often be expressed as a sum of components which, if commuting, allow the exponential to factorize. In the case where
$$M = M_a + M_b,$$
with M_a and M_b commuting (i.e., $M_a M_b = M_b M_a$), the equality
$$\exp(M) = \exp(M_a)\exp(M_b)$$
holds exactly. Such a decomposition is frequently encountered when M encapsulates both bivector and scalar parts. In many practical applications a multivector is decomposed into its even and odd grade parts:
$$M = M_{\text{even}} + M_{\text{odd}},$$
and under suitable conditions, the even part may admit a closed-form solution. The extraction of such closed forms is not only mathematically elegant but also provides computational leverage in simulation and modeling within physical systems.

Computational Considerations

The numerical evaluation of $\exp(M)$ is achieved by truncating the series expansion at a finite number of terms. In a computer algebra setting, multivectors are represented by sparse data structures where a basis blade (for example, denoted by a tuple) is mapped to its coefficient. The computation must address the evaluation of terms of the form M^n, which necessitates the implementation of an efficient and numerically stable geometric product. Careful management of floating-point arithmetic is required, as the convergence of the series is sensitive to the magnitude of the multivector. Additionally, techniques akin to scaling and squaring, familiar from the matrix exponential, may be adapted to reduce computation time while maintaining numerical accuracy. This involves scaling the multivector by an appropriate factor to ensure that its norm is sufficiently small, computing the exponential of the scaled multivector, and subsequently squaring the result a requisite number of times.

Python Implementation of a Multivector Exponential Function

A concise implementation of the exponential function for multivectors may be achieved by utilizing a power series expansion. The function below illustrates a single Python function that accepts a multivector represented as a dictionary and computes $\exp(M)$ by iteratively accumulating series terms. The auxiliary functions employed, such as those for performing the geometric product, scalar division, and addition of multivectors, are assumed to be defined elsewhere in the computational framework.

```
def exp_multivector(mv, n_terms=20):
    """
    Compute the exponential of a multivector 'mv' using its power
    ↪ series expansion.
        exp(mv) = sum_{n=0}^{infty} (mv^n) / n!

    Parameters:
        mv (dict): Dictionary representation of the multivector,
        ↪ where keys are basis blades.
        n_terms (int): Number of terms to include in the series
        ↪ expansion.

    Returns:
        dict: The multivector representing exp(mv).
    """
    result = {(): 1.0}   # Initialize with the scalar identity.
    term = {(): 1.0}     # The zeroth term in the series.
    for n in range(1, n_terms):
        term = multivector_multiply(term, mv)   # Compute successive
        ↪ powers mv^n.
        term = multivector_scalar_divide(term, math.factorial(n))
        result = multivector_add(result, term)
    return result
```

Python Code Snippet

```
import math
from copy import deepcopy

def sort_with_sign(lst):
    """
    Sort a list and return the sorted list along with the
    ↪ permutation sign.
    Uses a simple bubble sort algorithm.
```

```
    """
    l = lst[:]   # Create a copy
    sign = 1
    n = len(l)
    for i in range(n):
        for j in range(0, n - i - 1):
            if l[j] > l[j + 1]:
                l[j], l[j + 1] = l[j + 1], l[j]
                sign *= -1
    return l, sign

def geometric_product_blade(b1, b2):
    """
    Compute the geometric product of two basis blades b1 and b2.
    Each blade is represented as a tuple of integers, e.g., () for
    ↪ the scalar,
    (1,) for e1, (1,2) for e1^e2, etc.

    This function assumes an orthonormal Euclidean metric where the
    ↪ square of any basis vector is 1.
    It works by concatenating and then sorting the basis elements
    ↪ (tracking the permutation sign)
    and by canceling duplicate basis vectors since e_i * e_i = 1.

    Returns:
        (result_blade, factor)
    such that b1 * b2 = factor * result_blade,
    where result_blade is a sorted tuple with duplicates removed.
    """
    # Combine the two blades
    combined = list(b1) + list(b2)
    # Sort the combined list and extract the permutation sign
    sorted_combined, sign = sort_with_sign(combined)

    # Remove duplicate basis vectors (each duplicate pair cancels
    ↪ out with factor 1 in Euclidean space)
    result = []
    i = 0
    factor = 1.0
    while i < len(sorted_combined):
        if i < len(sorted_combined) - 1 and sorted_combined[i] ==
        ↪ sorted_combined[i + 1]:
            # e_i * e_i = 1 in an orthonormal Euclidean metric
            factor *= 1.0   # Here, metric(e_i)=1.0
            i += 2   # Skip the duplicate pair
        else:
            result.append(sorted_combined[i])
            i += 1
    return (tuple(result), sign * factor)

def multivector_add(mv1, mv2):
    """
    Add two multivectors represented as dictionaries.
```

```
        Each key is a basis blade (tuple) and each value is its
    ↪    coefficient.
    """
    result = deepcopy(mv1)
    for blade, coeff in mv2.items():
        result[blade] = result.get(blade, 0.0) + coeff
    return result

def multivector_scalar_divide(mv, scalar):
    """
    Divide a multivector by a scalar.
    """
    return {blade: coeff / scalar for blade, coeff in mv.items()}

def multivector_multiply(mv1, mv2):
    """
    Compute the geometric product of two multivectors.
    Each multivector is represented as a dictionary mapping basis
    ↪    blades (tuples) to coefficients.
    """
    result = {}
    for blade1, coeff1 in mv1.items():
        for blade2, coeff2 in mv2.items():
            prod_blade, factor = geometric_product_blade(blade1,
                ↪    blade2)
            coeff = coeff1 * coeff2 * factor
            result[prod_blade] = result.get(prod_blade, 0.0) + coeff
    return result

def multivector_power(mv, n):
    """
    Compute the nth power of a multivector using repeated geometric
    ↪    multiplication.
    """
    if n == 0:
        return {(): 1.0}  # Identity scalar
    result = deepcopy(mv)
    for _ in range(n - 1):
        result = multivector_multiply(result, mv)
    return result

def multivector_norm(mv):
    """
    Compute a simple norm of a multivector as the square root of the
    ↪    sum of squares of its coefficients.
    Note: This is not the full geometric norm but serves as a
    ↪    measure for scaling.
    """
    return math.sqrt(sum(coeff ** 2 for coeff in mv.values()))

def exp_multivector(mv, n_terms=20):
    """
```

```
Compute the exponential of a multivector 'mv' using its power
    series expansion with scaling and squaring.

The exponential is defined by:
    exp(mv) = sum_{n=0}^{infty} (mv^n) / n!

For numerical efficiency, the following steps are employed:
    1. Scaling: Determine an integer s such that the norm of
       mv/2^s is <= 1.
    2. Compute the exponential of the scaled multivector using the
       truncated power series.
    3. Squaring: Undo the scaling by squaring the result s times.

Parameters:
    mv (dict): Multivector represented as a dictionary.
    n_terms (int): Number of terms to include in the series
       expansion.

Returns:
    dict: The multivector representing exp(mv).
"""
# Scaling: reduce the norm of mv for improved series
    convergence.
norm = multivector_norm(mv)
s = 0
if norm > 1:
    s = math.ceil(math.log(norm, 2))
scaling_factor = 2 ** s
mv_scaled = multivector_scalar_divide(mv, scaling_factor)

# Compute exp(mv_scaled) using the power series:
# exp(mv_scaled) = sum_{n=0}^{n_terms-1} (mv_scaled^n) / n!
result = {(): 1.0}    # The zeroth term (scalar identity)
term = {(): 1.0}      # Initialize with mv_scaled^0 / 0! = 1
for n in range(1, n_terms):
    term = multivector_multiply(term, mv_scaled)
    term = multivector_scalar_divide(term, math.factorial(n))
    result = multivector_add(result, term)

# Squaring: Resume original scale by repeatedly squaring the
    series result.
for _ in range(s):
    result = multivector_multiply(result, result)
return result

# Example usage:
if __name__ == "__main__":
    # Define a bivector M represented as (1,2) with a coefficient.
    # In many GA applications, a bivector M satisfying M^2 = -² is
        used to represent rotations.
    # For our Euclidean GA, (1,2) normally squares to -1 when the
        proper sign convention is applied.
```

```python
# Here, we mimic this behavior by choosing a bivector with
↪    coefficient .
theta = 1.0  # Angle (or magnitude) parameter
M = {(1, 2): theta}  # Bivector representation

# Compute the exponential of the bivector M.
# The expected closed form for a bivector with M^2 = -² is:
#    exp(M) = cos() + (M/) sin()
exp_M = exp_multivector(M, n_terms=25)

# Display the computed multivector exponential.
print("Computed exp(M):")
for blade, coeff in exp_M.items():
    print(f"{blade}: {coeff}")
```

Chapter 15

Logarithms in GA: An Analytical Perspective

Foundations of Multivector Logarithms

The logarithmic function within geometric algebra extends the familiar scalar logarithm to the realm of multivectors, which may incorporate elements of various grades. For a multivector M, the logarithm, denoted by $\log(M)$, is formally defined as the inverse operation to the exponential map, so that

$$\exp\bigl(\log(M)\bigr) = M.$$

A common situation arises when the multivector can be decomposed into a scalar magnitude and a rotational component. In such cases, one may represent

$$M = \rho R,$$

where ρ is a positive real scalar and R is a rotor given by the exponential of a bivector, namely $R = \exp(B)$. Under these conditions, the logarithm admits a partitioning of the form

$$\log(M) = \log(\rho) + \log(R) = \log(\rho) + B,$$

with the understanding that the choice of branch for the logarithmic function must be handled with care.

Series Expansion and Convergence Considerations

When the multivector under consideration is in close proximity to the identity, it is often expressed as

$$M = I + X,$$

where I denotes the scalar identity and X is a multivector perturbation satisfying $\|X\| < 1$. In this regime, the power series expansion for the logarithm mirrors that of the scalar case:

$$\log(I + X) = \sum_{n=1}^{\infty} \frac{(-1)^{n+1}}{n} X^n.$$

This series converges provided that the norm of the deviation X remains within the radius of convergence. The evaluation of successive powers X^n requires careful handling of the geometric product, which is both associative and, in general, non-commutative. The alternating nature of the series confers advantages in terms of numerical stability when the condition $\|X\| < 1$ is maintained.

Branch Structure and Multivector Decomposition

For multivectors that are not near the identity, a decomposition into a norm and a directional component becomes essential. If a multivector can be written as

$$M = \rho R,$$

with $R = \exp(B)$ for a bivector B, then the logarithm divides naturally into

$$\log(M) = \log(\rho) + \log(R).$$

In many practical scenarios, the rotor R satisfies the condition that its logarithm is given by the corresponding bivector,

$$\log(R) = B,$$

where the bivector B represents the magnitude and axis of rotation. This decomposition is analogous to expressing a complex number in polar form and taking the logarithm by separating the modulus from the argument. The careful selection of the principal branch for the logarithm is required to ensure consistency of the inverse relationship with the exponential function.

Computational Strategies for Evaluating $\log(M)$

The computation of the logarithm for a general multivector is not straightforward due to the presence of multiple grades and the potential non-commutativity of the factors. A practical computational strategy involves scaling the multivector so that it is close to the identity. If a suitable integer k can be found such that the scaled multivector $M^{1/k}$ satisfies

$$M^{1/k} = I + X \quad \text{with} \quad \|X\| < 1,$$

then the logarithm may be computed via the power series expansion with

$$\log(M^{1/k}) = \sum_{n=1}^{\infty} \frac{(-1)^{n+1}}{n} X^n.$$

The original logarithm is recovered by the relation

$$\log(M) = k \log(M^{1/k}).$$

This scaling strategy is reminiscent of the techniques employed in matrix logarithms and is adapted here to respect the structure of geometric algebra. The intricate interplay of multivector grades calls for iterative methods and careful numerical management during the evaluation of the series.

Python Implementation of the Logarithmic Function

A computational implementation for evaluating the logarithm of a multivector uses the power series expansion approach under the assumption that the multivector can be cast in the form $M = I + X$ with $\|X\| < 1$. The function below illustrates a Python routine that computes $\log(M)$ using a dictionary representation of multivectors. This implementation presumes the availability of auxiliary functions for the subtraction, scalar multiplication, addition, and geometric product of multivectors.

```
def log_multivector(mv, n_terms=20):
    """
    Compute the logarithm of a multivector 'mv' using its power
    ↪ series expansion.
    Assumes that 'mv' can be represented in the form I + X, where I
    ↪ is the scalar
    identity and the norm of X is less than one, ensuring
    ↪ convergence of the series.

    The logarithm is given by:
        log(mv) = sum_{n=1}^{n_terms} ((-1)^(n+1)/n) * (X^n),
    where X = mv - I.

    Parameters:
        mv (dict): Dictionary representation of the multivector with
        ↪ basis blades as keys.
        n_terms (int): Number of terms in the expansion to compute.

    Returns:
        dict: The multivector representing log(mv).
    """
    I = {(): 1.0}
    X = multivector_subtract(mv, I)
    result = {}
    X_power = X
    for n in range(1, n_terms + 1):
        coeff = ((-1)**(n + 1)) / n
        term = multivector_scalar_multiply(X_power, coeff)
        result = multivector_add(result, term)
        X_power = multivector_multiply(X_power, X)
    return result
```

Python Code Snippet

```
# Multivector representation and operations in Geometric Algebra
# A multivector is represented as a dictionary where:
#   - keys are tuples representing basis blades (e.g., () for the
↪ scalar, (1,) for e1, (1,2) for e1^e2, etc.)
#   - values are the corresponding coefficients (real numbers)
#
# The following code implements:
#   1. Basic operations: addition, subtraction, scalar
↪ multiplication.
#   2. The geometric product between two multivectors.
#   3. A helper function to multiply a basis blade by a basis
↪ vector.
```

```
#   4. The computation of the multivector logarithm using the power
↪   series expansion.
#   5. A test routine that verifies the logarithm by computing an
↪   exponential,
#       analogous to the relationship exp(log(M)) = M.

def multiply_blade_vector(blade, v):
    """
    Multiply a basis blade by a basis vector.

    Parameters:
      blade (tuple): A sorted tuple representing the basis blade.
      v (int): An integer representing the basis vector (e.g., 1 for
      ↪ e1).

    Returns:
      tuple: (new_blade, sign) where new_blade is the resulting
      ↪ blade after multiplication,
             and sign is the factor induced by the reordering in the
             ↪ geometric product.

    The multiplication follows these rules in Euclidean space:
      - If the basis vector is not already in the blade, it is
      ↪ inserted in order.
        The sign factor is (-1)^p where p is the insertion position.
      - If the basis vector is already present, it is removed (since
      ↪ e_i*e_i = 1),
        and the sign factor is (-1)^p where p is the position where
        ↪ it was found.
    """
    new_blade = list(blade)
    # Loop through the current blade to find the insertion point or
    ↪ a duplicate.
    for i, b in enumerate(new_blade):
        if b == v:
            # Found duplicate: remove b and return with sign factor
            ↪ (-1)^i.
            del new_blade[i]
            return (tuple(new_blade), (-1)**i)
        elif b > v:
            # Insert v before the first larger element.
            new_blade.insert(i, v)
            return (tuple(new_blade), (-1)**i)
    # If v is larger than all existing elements, append it at the
    ↪ end.
    new_blade.append(v)
    return (tuple(new_blade), (-1)**(len(blade)))

def gp_blades(blade1, blade2):
    """
    Compute the geometric product of two basis blades.

    Parameters:
```

```
    blade1 (tuple): The first basis blade (sorted tuple).
    blade2 (tuple): The second basis blade (sorted tuple).

Returns:
    tuple: (result_blade, total_sign) where result_blade is the
    ↪ resulting basis blade
        after the product and total_sign is the overall sign
        ↪ from successive multiplications.
"""
result_blade = blade1
total_sign = 1
for v in blade2:
    result_blade, s = multiply_blade_vector(result_blade, v)
    total_sign *= s
return (result_blade, total_sign)

def multivector_add(mv1, mv2):
    """
    Add two multivectors.

    Parameters:
    mv1 (dict): First multivector.
    mv2 (dict): Second multivector.

    Returns:
    dict: The sum of mv1 and mv2.
    """
    result = mv1.copy()
    for blade, coeff in mv2.items():
        if blade in result:
            result[blade] += coeff
        else:
            result[blade] = coeff
    # Remove terms with negligible coefficients
    result = {blade: coeff for blade, coeff in result.items() if
    ↪ abs(coeff) > 1e-12}
    return result

def multivector_scalar_multiply(mv, scalar):
    """
    Multiply a multivector by a scalar.

    Parameters:
    mv (dict): Multivector.
    scalar (float): Scalar to multiply.

    Returns:
    dict: The resulting multivector.
    """
    return {blade: coeff * scalar for blade, coeff in mv.items()}

def multivector_subtract(mv1, mv2):
    """
```

    Subtract one multivector from another.

    Parameters:
      mv1 (dict): Minuend multivector.
      mv2 (dict): Subtrahend multivector.

    Returns:
      dict: The result of mv1 - mv2.
    """
    return multivector_add(mv1, multivector_scalar_multiply(mv2,
    ↪   -1))

def multivector_multiply(mv1, mv2):
    """
    Compute the geometric product of two multivectors.

    Parameters:
      mv1 (dict): First multivector.
      mv2 (dict): Second multivector.

    Returns:
      dict: The geometric product mv1 * mv2.
    """
    result = {}
    for blade1, coeff1 in mv1.items():
        for blade2, coeff2 in mv2.items():
            new_blade, sign = gp_blades(blade1, blade2)
            new_coeff = coeff1 * coeff2 * sign
            if new_blade in result:
                result[new_blade] += new_coeff
            else:
                result[new_blade] = new_coeff
    # Remove terms with negligible coefficients.
    result = {blade: coeff for blade, coeff in result.items() if
    ↪   abs(coeff) > 1e-12}
    return result

def log_multivector(mv, n_terms=20):
    """
    Compute the logarithm of a multivector 'mv' using a power series
    ↪   expansion.

    It is assumed that the multivector can be expressed as:
        mv = I + X,
    where I is the scalar identity and ||X|| < 1 (ensuring the
    ↪   convergence of the series).

    The logarithm is given by:
        log(mv) = sum_{n=1}^{n_terms} ((-1)^(n+1)/n) * X^n,
    with X = mv - I.

    Parameters:
      mv (dict): The input multivector in dictionary form.

```
        n_terms (int): Number of terms in the power series expansion.

    Returns:
        dict: The computed logarithm of mv.
    """
    # Define the identity multivector.
    I = {(): 1.0}
    # Compute X such that mv = I + X.
    X = multivector_subtract(mv, I)
    result = {}
    X_power = X
    for n in range(1, n_terms + 1):
        coeff = ((-1)**(n + 1)) / n
        term = multivector_scalar_multiply(X_power, coeff)
        result = multivector_add(result, term)
        X_power = multivector_multiply(X_power, X)
    return result

def exp_multivector(mv, n_terms=20):
    """
    Compute the exponential of a multivector using its power series
    ↪ expansion.

    The exponential is defined as:
        exp(mv) = I + mv + (1/2!) mv^2 + (1/3!) mv^3 + ...,
    where I is the scalar identity.

    Parameters:
        mv (dict): The input multivector.
        n_terms (int): Number of terms in the series expansion.

    Returns:
        dict: The multivector representing exp(mv).
    """
    from math import factorial
    I = {(): 1.0}
    result = I.copy()
    term = I.copy()
    for n in range(1, n_terms + 1):
        term = multivector_multiply(term, mv)
        term_scaled = multivector_scalar_multiply(term, 1.0 /
        ↪ factorial(n))
        result = multivector_add(result, term_scaled)
    return result

# Example usage: Compute the logarithm of a multivector and verify
↪ via the exponential.
if __name__ == "__main__":
    # Define a multivector M close to the identity:
    # Let M = I +  * e1, where  is small and e1 is represented by
    ↪ (1,).
    epsilon = 0.1
    M = {
```

119

```
    (): 1.0,       # Scalar part (identity)
    (1,): epsilon  # Small perturbation along e1
}

print("Input multivector M:")
for blade, coeff in sorted(M.items()):
    print(f"Blade {blade}: {coeff}")

# Compute the logarithm of M using series expansion.
logM = log_multivector(M, n_terms=20)

print("\nComputed log(M):")
for blade, coeff in sorted(logM.items()):
    print(f"Blade {blade}: {coeff}")

# Verify the inverse relationship by computing exp(log(M))
exp_logM = exp_multivector(logM, n_terms=20)

print("\nExponential of log(M) (should approximate M):")
for blade, coeff in sorted(exp_logM.items()):
    print(f"Blade {blade}: {coeff}")
```

Chapter 16

Rotors: Generators of Rotations in GA

Mathematical Framework of Rotors in Geometric Algebra

In geometric algebra, rotors are specialized multivectors that generate rotations in a given vector space. A rotor is typically defined as the exponential of a bivector, which encodes the plane and angle of rotation. In an orthonormal basis, a bivector B represents the oriented plane of rotation and its magnitude is proportional to the rotation angle. The rotor is constructed as

$$R = \exp\left(-\frac{B}{2}\right),$$

ensuring that the inverse of the rotor is given by its reversion,

$$R^{-1} = \tilde{R},$$

where the reversion operation reverses the order of basis elements in each blade. The rotation of a vector v is effected through the sandwiching operation,

$$v' = R v \tilde{R}.$$

This operation guarantees that the transformed vector retains its norm and that rotations compose in an associative manner under the geometric product.

Construction and Properties of Rotors

The construction of rotors follows directly from the exponential mapping within the algebra. Given a bivector B in the plane of rotation, the series expansion of the exponential yields

$$R = \exp\left(-\frac{B}{2}\right) = \cos\left(\frac{\theta}{2}\right) - \hat{B}\sin\left(\frac{\theta}{2}\right),$$

where θ is the rotation angle and $\hat{B}$ is the unit bivector corresponding to B. This formulation parallels the polar representation of complex numbers with cosine and sine functions appearing naturally in the expansion. The normalization condition,

$$R\tilde{R} = 1,$$

is guaranteed by the intrinsic properties of the exponential map when applied to bivectors. The careful handling of scalar and bivector parts in the series is imperative for both theoretical insights and numerical implementations.

Implementation in Computational Frameworks

The practical application of rotors in computational environments necessitates a precise handling of multivector operations, especially the geometric product and the reversion operation. Efficient rotor application involves computing the expression

$$v' = Rv\tilde{R},$$

which for computational purposes is implemented through optimized routines for multivector multiplication and reverse evaluation. The implementation must account for the non-commutative nature of the geometric product and preserve numerical stability across operations. An essential function in this context is the application of a rotor to a vector, which encapsulates the core routine of rotor-based transformations.

Python Implementation of Rotor Application

The function below demonstrates a Python routine that applies a rotor to a vector. In this implementation, multivectors are represented as dictionaries, and the rotor is assumed to be normalized such that its reverse computes efficiently. The function multiplies the rotor by the vector and then by the reverse of the rotor, yielding the rotated vector.

```
def apply_rotor(vector, rotor):
    """
    Compute the rotated vector using a rotor in geometric algebra.

    The rotation is executed by the transformation:
      v_rotated = rotor * vector * reverse(rotor)
    where 'reverse(rotor)' denotes the reversion of the rotor.

    Parameters:
      vector (dict): Dictionary representing the vector in geometric
        ↪ algebra.
      rotor (dict): Dictionary representing the rotor (normalized
        ↪ multivector).

    Returns:
      dict: The rotated vector.
    """
    return multivector_multiply(
        multivector_multiply(rotor, vector),
        reverse(rotor)
    )
```

Python Code Snippet

```
import math

def blade_product(blade1, blade2):
    """
    Compute the geometric product of two basis blades.
    Each blade is represented as a tuple of integers in sorted
      ↪ order.
    The function returns a tuple (result_blade, sign) where:
      - result_blade is the sorted tuple representing the resulting
        ↪ blade,
      - sign is the accumulated sign from swapping basis vectors.
    """
```

```python
    result = list(blade1)
    sign = 1
    for b in blade2:
        # Count the number of elements in result greater than b
        ↪ (each swap introduces a -1)
        swaps = sum(1 for a in result if a > b)
        sign *= (-1) ** swaps
        if b in result:
            # If the same basis vector appears, it squares to 1 in
            ↪ Euclidean space and cancels.
            result.remove(b)
        else:
            result.append(b)
            result.sort()
    return tuple(result), sign

def gp(mv1, mv2):
    """
    Compute the geometric product of two multivectors.
    A multivector is represented as a dictionary with:
      keys: tuples representing basis blades (e.g., () for scalar,
      ↪ (1,) for e1, (1,2) for e12, etc.)
      values: coefficients (floats)
    The product is computed by summing over the products of each
    ↪ pair of basis blades.
    """
    result = {}
    for blade1, coeff1 in mv1.items():
        for blade2, coeff2 in mv2.items():
            new_blade, sign = blade_product(blade1, blade2)
            prod = coeff1 * coeff2 * sign
            if new_blade in result:
                result[new_blade] += prod
            else:
                result[new_blade] = prod
    # Remove terms with negligible coefficients.
    result = {blade: coeff for blade, coeff in result.items() if
    ↪ abs(coeff) > 1e-12}
    return result

def reverse(mv):
    """
    Compute the reversion (reverse) of a multivector.
    Reversion reverses the order of basis vectors in every blade.
    For a blade of grade k, the reversion factor is
    ↪ (-1)^(k*(k-1)/2).
    """
    result = {}
    for blade, coeff in mv.items():
        grade = len(blade)
        factor = (-1) ** (grade * (grade - 1) // 2)
        result[blade] = coeff * factor
    return result
```

```python
def scalar_mult(mv, scalar):
    """
    Multiply every component of a multivector by a scalar.
    """
    return {blade: coeff * scalar for blade, coeff in mv.items()}

def add_mv(mv1, mv2):
    """
    Add two multivectors.
    """
    result = mv1.copy()
    for blade, coeff in mv2.items():
        if blade in result:
            result[blade] += coeff
        else:
            result[blade] = coeff
    # Remove near-zero coefficients.
    result = {blade: coeff for blade, coeff in result.items() if
    ↪   abs(coeff) > 1e-12}
    return result

def bivector_square(B):
    """
    Compute the square of a bivector B using the geometric product.
    For a simple bivector in Euclidean space, $B^2$ is a scalar equal
    ↪   to $-|B|^2$.
    """
    B2 = gp(B, B)
    return B2.get((), 0.0)

def exp_bivector(B):
    """
    Compute the exponential of a bivector to generate a rotor.
    For a simple bivector B, the exponential mapping is given by:
        exp(-B/2) = cos(|B|/2) - hat(B)*sin(|B|/2)
    where |B| is the magnitude of the bivector and hat(B) is its
    ↪   normalized version.
    """
    # Compute the square of B. In Euclidean GA, $B^2$ = $-|B|^2$ for a
    ↪   simple bivector.
    B2 = bivector_square(B)
    normB = math.sqrt(-B2) if B2 < 0 else 0
    if normB == 0:
        # When B is zero, return the identity rotor.
        return {(): 1.0}
    # Normalize B to obtain the unit bivector hat(B).
    hatB = scalar_mult(B, 1.0 / normB)
    cos_term = math.cos(normB / 2)
    sin_term = math.sin(normB / 2)
    # Construct the rotor: R = cos(normB/2) - hat(B)*sin(normB/2)
    rotor = add_mv({(): cos_term}, scalar_mult(hatB, -sin_term))
    return rotor
```

```python
def apply_rotor(vector, rotor):
    """
    Rotate a vector using the rotor via the sandwiching operation:
        v' = rotor * vector * reverse(rotor)
    """
    return gp(gp(rotor, vector), reverse(rotor))

# Example usage:
if __name__ == "__main__":
    # Define basis blades for 3D geometric algebra:
    # Scalar: ()
    # Vectors: (1,) represents e1, (2,) represents e2, (3,) represents e3.
    # Bivectors: (1,2) represents e12, etc.

    # Define a bivector representing a rotation in the e1-e2 plane.
    # For a rotation by 45 degrees, let theta be in radians.
    theta = math.radians(45)  # 45 degrees in radians.
    # Bivector B = theta * e12 (i.e., B = theta*(e1^e2))
    B = {(1, 2): theta}

    # Compute the rotor R = exp(-B/2)
    rotor = exp_bivector(B)

    # Define a vector v along the e1 direction.
    v = {(1,): 1.0}

    # Apply the rotor to the vector.
    v_rotated = apply_rotor(v, rotor)

    # Output the rotor and the rotated vector.
    print("Rotor:")
    for blade, coeff in sorted(rotor.items()):
        print(f"{blade}: {coeff}")

    print("\nOriginal vector:")
    print(v)

    print("\nRotated vector:")
    for blade, coeff in sorted(v_rotated.items()):
        print(f"{blade}: {coeff}")
```

Chapter 17

Reflections: Representation through Geometric Algebra

Mathematical Framework of Reflections in GA

Reflections constitute a fundamental transformation in geometric algebra, encapsulating the inversion of a vector across a hyperplane. When a hyperplane is specified by a unit normal vector n, the reflection of an arbitrary vector v is achieved through a sandwiching operation. In the algebraic framework, the reflection is expressed as

$$v' = -n\,v\,n,$$

where the geometric product intrinsically combines both the inner and outer products. This representation adheres to the principle that the geometric product of a unit vector with itself satisfies $n^2 = 1$, and the negative sign naturally corrects the orientation by inverting the component of v normal to the hyperplane. This coordinate-free description lays the groundwork for both analytical derivation and computational modeling of reflections.

Derivation of Reflection Formulas

The derivation of the reflection formula commences by decomposing a vector v into two components: one parallel to the hyperplane, denoted by $v_\perp$, and one perpendicular to it, denoted by $v_\parallel$. Given that
$$v_\parallel = (v \cdot n)n \quad \text{and} \quad v_\perp = v - (v \cdot n)n,$$
the classical reflection of the vector is written as
$$v' = v_\perp - v_\parallel = v - 2(v \cdot n)n.$$

Within the framework of geometric algebra, this same result is encapsulated by the sandwich transformation
$$v' = -n\,v\,n.$$

Expanding the product and applying the properties of the geometric product confirms that the sandwich form accurately reverses the sign of the component of v that is parallel to n while leaving the perpendicular component unaltered. The derivation emphasizes that the algebraic structure naturally incorporates the geometric intuition behind reflections.

Physical Interpretations of Reflection Operations

The reflection operation in geometric algebra bears significant physical connotations. In optics, for instance, reflections describe how incident light vectors are inverted when encountering a reflective surface, encapsulated by the transformation of electric field vectors. The operation $v' = -n\,v\,n$ provides a rigorous, coordinate-independent method to simulate such phenomena. In the realm of mechanics and quantum theory, reflections implement discrete symmetry operations that are central to analyses of parity and inversion transformations. The algebraic formalism not only streamlines the derivation of reflection formulas but also reveals the intrinsic connection between geometric transformations and physical laws, allowing for a unified treatment of symmetry operations across different physical systems.

Python Implementation: Reflection Function

A concise computational implementation of the reflection operation is achieved by defining a function that performs the sandwich product with a unit normal vector. The function below encapsulates the operation by computing

$$v_{\text{reflected}} = -n\,v\,n,$$

assuming that both the vector and the normal are represented as multivectors and that the normal vector is pre-normalized.

```python
def reflect_vector(vector, normal):
    """
    Reflects a vector across a hyperplane defined by a normalized
    ↪ unit normal.

    The reflection is implemented using the sandwich product:
        v_reflected = - normal * vector * normal

    Parameters:
        vector (dict): Multivector representation of the vector to be
        ↪ reflected.
        normal (dict): Multivector representation of the normalized
        ↪ unit normal.

    Returns:
        dict: Multivector representation of the reflected vector.
    """
    return multivector_multiply(
        multivector_multiply(scalar_mult(normal, -1), vector),
        normal
    )
```

Python Code Snippet

```python
import numpy as np

def dot(v, w):
    """
    Compute the dot product of two vectors v and w.
    Both v and w are expected to be numpy arrays.
    """
    return np.dot(v, w)
```

```
def multivector_reflection(v, n):
    """
    Reflect a vector v across a hyperplane defined by a unit normal
    ↪ vector n
    using the geometric algebra sandwich product representation:

        v' = - n * v * n

    For vectors in a Euclidean space and for a normalized n (i.e.
    ↪ n^2 = 1),
    this operation is equivalent to the classical reflection
    ↪ formula:

        v' = v - 2 * (v dot n) * n

    Parameters:
      v (numpy.array): The vector to be reflected.
      n (numpy.array): The unit normal vector of the hyperplane.

    Returns:
      numpy.array: The reflected vector.
    """
    # In practice, Python does not natively implement
    ↪ non-commutative geometric products.
    # However, by using the fact that n is normalized, the sandwich
    ↪ product reduces to:
    return v - 2 * dot(v, n) * n

def reflect_vector(v, n):
    """
    Demonstrate the computation of the reflection of a vector using
    ↪ the GA sandwich
    product and the equivalent classical reflection method.

    In geometric algebra the reflection is given by:
        v' = - n * v * n
    which, for a normalized n, is equivalent to:
        v' = v - 2 * (v dot n) * n

    Parameters:
      v (numpy.array): The vector to be reflected.
      n (numpy.array): The normalized unit normal vector defining
      ↪ the hyperplane.

    Returns:
      dict: A dictionary containing the reflected vector computed
      ↪ via the
            GA sandwich product ('sandwich') and the classical
            ↪ formula ('classical').
    """
    # Since we do not have a dedicated GA algebra package here, we
    ↪ simulate the
    # sandwich product using the classical reflection equation.
```

```python
    reflected = multivector_reflection(v, n)

    return {
        'sandwich': reflected,
        'classical': reflected
    }

def main():
    """
    Main routine to demonstrate the reflection operation in
      geometric algebra.

    A sample 3D vector and a unit normal vector are defined. The
      vector is then
    reflected across the corresponding hyperplane. For a vector v
      and unit normal n,
    the reflection is computed via:

        v' = v - 2 * (v dot n) * n

    For example, reflecting a vector across the xz-plane (normal
      along y) should
    invert its y-component.
    """
    # Define a sample vector v in 3D space.
    v = np.array([3.0, 4.0, 0.0])

    # Define the hyperplane's unit normal vector n.
    # For instance, choose n along the y-axis so that the xz-plane
    #   is the reflecting plane.
    n = np.array([0.0, 1.0, 0.0])
    n = n / np.linalg.norm(n)   # Ensure n is normalized

    # Compute the reflection using the GA sandwich product
    #   (simulated)
    result = reflect_vector(v, n)

    print("Original Vector: ", v)
    print("Unit Normal Vector: ", n)
    print("Reflected Vector (using sandwich product): ",
      result['sandwich'])
    print("Reflected Vector (using classical formula): ",
      result['classical'])
    # Expected output: For v = [3,4,0] and n = [0,1,0],
    # the reflected vector should be [3, -4, 0]

if __name__ == "__main__":
    main()
```

Chapter 18

Spinors: Their Role in Geometric Physics

Mathematical Foundations of Spinors

Spinors emerge naturally within the framework of geometric algebra as objects that encapsulate both rotation and intrinsic spin properties. In this context, a spinor is an element of the even subalgebra, formed by the geometric product of an even number of unit vectors. A typical construction expresses a spinor R in the exponential form

$$R = \exp\left(-\frac{B}{2}\right),$$

where B is a bivector that characterizes the plane of rotation. The normalization condition

$$R\tilde{R} = 1,$$

with $\tilde{R}$ denoting the reversion (i.e., the grade-reversal) of R, underlies the spinor's ability to represent rotations in a double-cover of the special orthogonal group. This formulation provides a coordinate-free and algebraically compact representation of rotations, which is a cornerstone in both classical physics and quantum theory.

Spinors and Rotational Symmetries

Within the geometric algebra framework, spinors serve as the generators of rotations via the sandwich product. Given a vector v, its rotation is effected by a spinor R according to

$$v' = Rv\tilde{R}.$$

This expression encodes the operation's inherent double-cover property; specifically, a rotation by an angle of 2π yields a sign inversion of the corresponding spinor while leaving the vector unchanged. The capacity of spinors to capture this subtle feature is pivotal in the treatment of particles with half-integer spin, where rotational symmetries are intimately linked to observable physical phenomena. The algebraic structure thereby bridges the formulation of intrinsic spin in quantum mechanics with the geometric representations of rotations.

Computational Approach to Spinor Rotations

The computational realization of spinor rotations leverages the concise structure of the sandwich product. In numerical simulations, a spinor is typically derived from a bivector encoding both the plane of rotation and the rotation angle. The rotation of a vector v is computed via

$$v' = Rv\tilde{R},$$

ensuring that the unit norm condition $R\tilde{R} = 1$ is maintained throughout the calculation. The following Python function implements the spinor rotation of a vector, assuming that the operations `multivector_multiply` and `reverse` are defined within the computational framework to handle the geometric product and reversion, respectively.

```
def rotate_vector_by_spinor(vector, spinor):
    """
    Apply a spinor rotation to a vector using the geometric algebra
    ↪ sandwich product.

    The rotated vector is computed as:
        rotated_vector = spinor * vector * reverse(spinor)
    where reverse(spinor) denotes the reversion (conjugation) of the
    ↪ spinor.

    Parameters:
      vector (dict or array): The representation of the vector to be
      ↪ rotated.
```

```
spinor (dict or array): The spinor representing the rotation;
↪    must satisfy spinor * reverse(spinor) = 1.

Returns:
    dict or array: The rotated vector.
"""
rotated_vector = multivector_multiply(
    multivector_multiply(spinor, vector),
    reverse(spinor)
)
return rotated_vector
```

Spinor Algebra in the Geometric Framework

The algebraic properties of spinors are central to their role in representing rotational symmetries. The non-commutative nature of the geometric product, combined with the reversion operation, ensures that spinors retain the necessary information to encode rotational transformations. When successive rotations are applied, the composition of the corresponding spinors (via spinor multiplication) yields a new spinor that represents the overall transformation. This inherent composability simplifies operations in systems where multiple rotations occur sequentially. Moreover, the double-cover property—whereby the spinor representation distinguishes between a rotation by 2π and the identity operation—provides a natural explanation for phenomena observed in quantum mechanics, particularly in the behavior of spin-$\frac{1}{2}$ particles.

The formulation of spinors as elements of the even subalgebra of geometric algebra unites the description of classical rotations with the sophisticated algebra of quantum spin. This unification creates a robust framework that permits both analytical derivations and computational implementations of physical transformations while preserving the underlying symmetry principles intrinsic to geometric physics.

Python Code Snippet

```
# Comprehensive Python implementation for geometric algebra
↪    operations
# including geometric product, reversion, and spinor rotation.
```

```python
import math
from bisect import bisect_left

# Helper function: Multiply a basis blade (an ordered tuple of basis
# ↪ indices)
# by a single basis vector (represented as an integer).
def blade_mult(blade, e):
    """
    Multiply a blade (represented as an ordered tuple of basis
    ↪ indices)
    by a basis vector 'e'. If 'e' is already present in the blade,
    ↪ it is removed
    (since e*e = 1 in Euclidean metric) with an appropriate sign
    ↪ determined by
    its position. Otherwise, 'e' is inserted in order with a sign
    ↪ corresponding
    to the number of swaps required.

    Parameters:
        blade (tuple): An ordered tuple representing a basis blade.
        e (int): The basis vector to multiply by.

    Returns:
        (tuple, int): A tuple containing the resultant blade (ordered
        ↪ tuple)
        and the sign factor (+1 or -1).
    """
    new_blade = list(blade)
    if e in new_blade:
        index = new_blade.index(e)
        del new_blade[index]
        return (tuple(new_blade), (-1)**index)
    else:
        # Insert 'e' maintaining the sorted order.
        pos = bisect_left(new_blade, e)
        new_blade.insert(pos, e)
        # The sign factor is (-1) raised to (number of elements
        ↪ shifted),
        # which is (len(blade) - pos).
        return (tuple(new_blade), (-1)**(len(blade) - pos))

# Multiply two basis blades (each represented as a tuple).
def multiply_blades(blade1, blade2):
    """
    Multiply two basis blades using the rules of the geometric
    ↪ product.
    The blades are represented as ordered tuples of basis indices.

    Parameters:
        blade1 (tuple): The first blade.
        blade2 (tuple): The second blade.

    Returns:
```

```
        (tuple, int): A tuple containing the resulting blade (as an
            ordered tuple)
        and the overall sign from the multiplication.
    """
    result_blade = blade1
    overall_sign = 1
    for e in blade2:
        result_blade, s = blade_mult(result_blade, e)
        overall_sign *= s
    return (result_blade, overall_sign)

# Geometric product between two multivectors.
# A multivector is represented as a dictionary mapping basis blades
    (tuples)
# to their scalar coefficients.
def multivector_multiply(mv1, mv2):
    """
    Compute the geometric product of two multivectors.

    Parameters:
        mv1 (dict): First multivector, with keys as basis blade tuples
            and values as coefficients.
        mv2 (dict): Second multivector, in the same format.

    Returns:
        dict: The resulting multivector from the geometric product.
    """
    result = {}
    for blade1, coeff1 in mv1.items():
        for blade2, coeff2 in mv2.items():
            prod_blade, sign = multiply_blades(blade1, blade2)
            coeff = coeff1 * coeff2 * sign
            if prod_blade in result:
                result[prod_blade] += coeff
            else:
                result[prod_blade] = coeff
    # Remove near-zero coefficients
    result = {blade: val for blade, val in result.items() if
        abs(val) > 1e-12}
    return result

# The reverse (or reversion) operation in geometric algebra.
# For a blade of grade r, the reversion multiplies the coefficient
    by (-1)^(r*(r-1)/2).
def reverse(mv):
    """
    Compute the reversion (grade reversal) of a multivector.

    Parameters:
        mv (dict): A multivector with keys as basis blade tuples.

    Returns:
        dict: The reversed multivector.
```

```python
    """
    rev = {}
    for blade, coeff in mv.items():
        grade = len(blade)
        sign = (-1)**(grade*(grade-1)//2)
        # Although the reversal formally reverses the order of the
        ↪ basis vectors,
        # since our blades are stored in sorted order, we simply
        ↪ incorporate the sign.
        rev[blade] = coeff * sign
    return rev

# Compute a rotor (spinor) from a simple bivector.
def rotor_from_bivector(B):
    """
    Given a simple bivector B (represented as a multivector with a
    ↪ single bivector term),
    compute the rotor (spinor) R = exp(-B/2) using the closed-form
    ↪ formula:
        R = cos(|B|/2) - (B/|B|)*sin(|B|/2)

    This rotor can then be used to effect rotations via the sandwich
    ↪ product.

    Parameters:
        B (dict): A bivector represented as a dictionary with one key
        ↪ of length 2.

    Returns:
        dict: The rotor (spinor) as a multivector.
    """
    if len(B) != 1:
        raise ValueError("Only simple bivectors are supported for
        ↪ rotor computation.")
    blade, b_val = list(B.items())[0]
    if len(blade) != 2:
        raise ValueError("Provided multivector is not a bivector.")
    theta = abs(b_val)   # The magnitude of the rotation angle.
    cos_val = math.cos(theta/2)
    sin_val = math.sin(theta/2)
    rotor = {}
    rotor[()] = cos_val   # Scalar part.
    # Bivector part (normalized) multiplied by -sin(theta/2).
    rotor[blade] = - (b_val/theta) * sin_val if theta != 0 else 0
    return rotor

# Rotate a vector using the spinor (rotor) via the sandwich product.
def rotate_vector_by_spinor(vector, spinor):
    """
    Rotate a vector 'vector' using the spinor (rotor) 'spinor'
    ↪ according to:
        v' = spinor * vector * reverse(spinor)
```

```
    Parameters:
        vector (dict): The vector to be rotated. Represented as a
        ↪  multivector with 1-blade keys.
        spinor (dict): The spinor (rotor) effecting the rotation.
        ↪  Should satisfy spinor * reverse(spinor) = 1.

    Returns:
        dict: The rotated vector as a multivector.
    """
    intermediate = multivector_multiply(spinor, vector)
    return multivector_multiply(intermediate, reverse(spinor))

# Example usage and test.
if __name__ == "__main__":
    # Define a vector v along the x-axis: v = e1.
    v = {
        (1,): 1.0,
        (2,): 0.0,
        (3,): 0.0
    }

    # Define a bivector for rotation in the e1^e2 plane.
    # For a rotation by 90 degrees (pi/2), set B = (pi/2) * (e1^e2).
    phi = math.pi / 2  # Rotation angle.
    B = {
        (1, 2): phi
    }

    # Compute the rotor (spinor) from the bivector B.
    R = rotor_from_bivector(B)

    # Rotate the vector v using the rotor R.
    v_rotated = rotate_vector_by_spinor(v, R)

    # Display the results.
    print("Original vector v:")
    for blade, coeff in v.items():
        print(f"  {blade}: {coeff:.6f}")

    print("\nRotor R (spinor):")
    for blade, coeff in R.items():
        print(f"  {blade}: {coeff:.6f}")

    print("\nRotated vector v':")
    for blade, coeff in v_rotated.items():
        print(f"  {blade}: {coeff:.6f}")
```

Chapter 19

Rotations in GA: From Theory to Computation

Fundamental Principles of Rotations in Geometric Algebra

Rotations in geometric algebra are expressed through the elegant construction of rotors, elements of the even subalgebra that enact transformations via the sandwich product. In this formulation, a rotor R is defined so as to satisfy the normalization condition

$$R\tilde{R} = 1,$$

where $\tilde{R}$ denotes the reversion (grade-reversal) of R. The action of a rotor on a vector v is given by

$$v' = Rv\tilde{R}.$$

The structure of this operation encapsulates the double-cover property inherent to the spin representation; a rotation by an angle of 2π returns the original vector while inverting the sign of the rotor. This feature, which distinguishes half-integer spin systems in quantum mechanics, is deeply instrumental in both physical modeling and computational simulations.

A rotor is often derived by exponentiating a bivector B, so that

$$R = \exp\left(-\frac{B}{2}\right).$$

Here, the bivector B encodes the plane of rotation as well as the rotation magnitude. The above exponential formulation provides a coordinate-free method to represent rotations and ensures that all trigonometric functions naturally appear in the final computations. The algebraic closure of the rotor under multiplication allows the composition of rotational transformations by simple rotor products, directly mirroring the non-commutative behavior of the underlying geometric product.

Computational Implementation of Rotational Transformations

The translation of these theoretical insights into a computational framework requires careful formulation of the geometric product and reversal operations. In numerical routines, one typically represents multivectors as data structures that map basis blades to scalar coefficients. The rotor is then applied to a vector by sequentially computing the product Rv and subsequently multiplying by $\tilde{R}$. This sandwich product operation has the computational advantage of preserving the norm and enacting the rotation in a compact form.

A single, well-encapsulated function can perform the rotation by taking as input a vector and a unit rotor and returning the rotated vector. The following Python function illustrates this operation. The function assumes that auxiliary operations, such as the geometric product function `multivector_multiply` and the reversion function `reverse`, are defined within the computational framework.

```
def rotate_vector_by_rotor(vector, rotor):
    """
    Compute the rotated vector using the geometric algebra sandwich
    ↪ product.

    The rotation operation is performed according to:
        v' = rotor * vector * reverse(rotor)
    The function assumes that the rotor is a unit multivector,
    ↪ meaning that:
        rotor * reverse(rotor) = 1.

    Parameters:
        vector (object): The representation of the vector to be
        ↪ rotated.
        rotor  (object): A unit rotor encoding the desired rotation.
```

```
    Returns:
        object: The rotated vector computed from the sandwich product.
    """
    return multivector_multiply(
            multivector_multiply(rotor, vector),
            reverse(rotor))
```

Within this function, the geometric product is used recursively to first combine the rotor with the vector and then to recombine the intermediate product with the rotor's reversion. Such a modular approach is conducive to efficient computational implementations, especially when integrated into larger simulation frameworks.

The precision of the rotation operation depends on both the numerical stability of the geometric product evaluation and the exact preservation of the unit norm for the rotor. Consequently, ensuring that each intermediary computational step adheres to the algebraic constraints is critical. Techniques such as re-normalization of the rotor or the adaptive management of floating point accuracy may be introduced in a broader software context, though the fundamental strategy remains rooted in the sandwich product formulation.

The seamless integration of theoretical geometric algebra with computational algorithms provides a robust framework for simulating physical rotations. The direct mapping of algebraic operations to computational functions not only reinforces the conceptual underpinnings but also enhances the practical applicability in solving complex problems within physics and engineering.

Python Code Snippet

```
import math

def geometric_product_basis(b1, b2):
    """
    Compute the geometric product of two basis blades.
    Each basis blade is represented as a sorted tuple of integers.
    The product is computed by sequentially inserting elements from
    ↪  b2 into b1,
    canceling duplicate elements (since e_i^2 = 1) and accumulating
    ↪  sign changes
    from swaps required to maintain sorted order.

    Parameters:
        b1 (tuple): A tuple representing the first basis blade.
        b2 (tuple): A tuple representing the second basis blade.
```

    Returns:
      tuple: A pair (result_blade, sign) where result_blade is the
      ↪ resulting basis
            blade (as a sorted tuple) and sign is +1 or -1
            ↪ depending on the reordering.
    """
    res = list(b1) # Start with the elements of the first blade.
    sign = 1
    for idx in b2:
        if idx in res:
            # When the same basis vector appears twice, it squares
            ↪ to 1, so cancel it.
            res.remove(idx)
        else:
            # Count the number of elements in res that are greater
            ↪ than idx.
            # Each such element implies a swap (and hence a sign
            ↪ change) when inserting.
            count = sum(1 for r in res if r > idx)
            if count % 2 != 0:
                sign *= -1
            # Insert idx into the list in sorted order.
            inserted = False
            for i, r in enumerate(res):
                if idx < r:
                    res.insert(i, idx)
                    inserted = True
                    break
            if not inserted:
                res.append(idx)
    return (tuple(res), sign)

def multivector_multiply(A, B):
    """
    Multiply two multivectors A and B.
    Both multivectors are represented as dictionaries mapping basis
    ↪ blades
    (expressed as sorted tuples) to their scalar coefficients.

    Parameters:
      A (dict): The first multivector.
      B (dict): The second multivector.

    Returns:
      dict: The resulting multivector after multiplication.
    """
    result = {}
    for blade1, coeff1 in A.items():
        for blade2, coeff2 in B.items():
            blade, sign = geometric_product_basis(blade1, blade2)
            coeff = coeff1 * coeff2 * sign
            if blade in result:

```
                    result[blade] += coeff
                else:
                    result[blade] = coeff
    # Eliminate coefficients that are numerically negligible.
    result = {blade: coeff for blade, coeff in result.items() if
    ↪   abs(coeff) > 1e-12}
    return result

def reverse(M):
    """
    Compute the reversion (grade-reversal) of a multivector M.
    For a basis blade of grade k, the reverse is given by
    ↪   multiplying the coefficient by (-1)^(k(k-1)/2).

    Parameters:
      M (dict): The multivector to reverse.

    Returns:
      dict: The reversed multivector.
    """
    result = {}
    for blade, coeff in M.items():
        grade = len(blade)
        factor = 1 if ((grade * (grade - 1) // 2) % 2 == 0) else -1
        result[blade] = coeff * factor
    return result

def bivector_exp(B):
    """
    Compute the exponential of a bivector B to form a rotor.
    This function assumes that B is a pure bivector (i.e., it
    ↪   contains only grade-2 components).
    With the Euclidean signature, the square of a bivector is
    ↪   negative and equals -|B|^2.
    Hence, exp(-B/2) can be computed as:
        R = cos(|B|/2) - (B/|B|) * sin(|B|/2)

    Parameters:
      B (dict): The bivector (as a multivector) whose exponential is
      ↪   to be computed.

    Returns:
      dict: The rotor obtained from the bivector exponential.
    """
    B2 = multivector_multiply(B, B)
    # Extract the scalar part of B^2.
    scalar_part = B2.get((), 0)
    # For a bivector in Euclidean space, B^2 should be negative.
    if scalar_part < 0:
        magnitude = math.sqrt(-scalar_part)
    else:
        magnitude = math.sqrt(scalar_part)
```

```python
        half = 0.5
        cos_term = math.cos(magnitude * half)
        sin_term = math.sin(magnitude * half)

        # Initialize the rotor with the scalar component.
        rotor = {(): cos_term}
        if abs(magnitude) > 1e-12:
            factor = -sin_term / magnitude
            for blade, coeff in B.items():
                rotor[blade] = rotor.get(blade, 0) + coeff * factor
        else:
            # When the magnitude is extremely small, no rotation is
            ↪ applied.
            for blade, coeff in B.items():
                rotor[blade] = rotor.get(blade, 0) + 0
        return rotor

def rotate_vector_by_rotor(vector, rotor):
    """
    Rotate a vector using a rotor via the geometric sandwich
    ↪ product.
    The rotated vector is computed as:
        v' = rotor * vector * reverse(rotor)

    Parameters:
        vector (dict): The vector to be rotated (should have grade 1
        ↪ only).
        rotor (dict): A unit rotor representing the rotation.

    Returns:
        dict: The rotated vector.
    """
    temp = multivector_multiply(rotor, vector)
    rotated = multivector_multiply(temp, reverse(rotor))
    return rotated

def print_multivector(M):
    """
    Utility function to print a multivector in a human-readable
    ↪ format.
    Each basis blade is printed with its coefficient.

    Parameters:
        M (dict): The multivector to print.
    """
    components = []
    for blade, coeff in M.items():
        if blade == ():
            basis = "1"
        else:
            basis = "e" + "".join(str(i) for i in blade)
        components.append(f"{coeff:+.4f}{basis}")
    print(" ".join(components))
```

```python
# Example usage demonstrating rotational transformation:
if __name__ == "__main__":
    # Define the basis vectors in 3D Euclidean space.
    e1 = {(1,): 1.0}
    e2 = {(2,): 1.0}
    e3 = {(3,): 1.0}

    # Define a vector to be rotated. For example, take v = e1.
    vector = e1.copy()
    print("Original vector:")
    print_multivector(vector)

    # Define a bivector representing the rotation plane.
    # For a rotation in the e1-e2 plane by 90 degrees (pi/2), the
    ↪  bivector is:
    angle = math.pi / 2  # 90 degrees rotation
    bivector = {(1, 2): angle}

    # Compute the rotor from the bivector using the exponential map:
    #     rotor = exp(-bivector/2)
    rotor = bivector_exp(bivector)
    print("\nRotor:")
    print_multivector(rotor)

    # Apply the rotor to rotate the vector via the sandwich product.
    rotated_vector = rotate_vector_by_rotor(vector, rotor)
    print("\nRotated vector:")
    print_multivector(rotated_vector)
```

Chapter 20

Lorentz Transformations: GA Approach to Relativity

Foundations of Lorentz Transformations in Geometric Algebra

In the relativistic context the Lorentz transformations preserve the Minkowski metric and mix temporal and spatial components of four-vectors. Geometric algebra provides a unified framework in which both rotations and boosts are represented by rotors. In this formalism the boost along a specified spatial direction is generated by the exponential of a bivector that spans a plane defined by a timelike vector and the chosen spatial direction. For example, for a boost in the t–x plane the bivector is constructed from the basis vectors γ_0 and γ_1, and the corresponding rotor is written as

$$R = \exp\left(-\frac{\phi}{2}\hat{B}\right),$$

where ϕ is the rapidity parameter and $\hat{B}$ is the normalized bivector. The Lorentz transformation of any multivector v is then effected via the sandwich product

$$v' = R\,v\,\tilde{R},$$

with $\tilde{R}$ denoting the reversion of R. This construction not only preserves the Minkowski norm but also reflects the double-cover property inherent in the spin representation of the Lorentz group.

Exponential Map for Boosts and Hyperbolic Dynamics

The exponential mapping of a bivector in Minkowski spacetime differs from its Euclidean counterpart in that the square of a boost bivector is positive. Consequently, the expansion of the exponential involves hyperbolic trigonometric functions rather than the familiar sine and cosine. Explicitly, if B is a bivector associated with a boost and its magnitude is given by ϕ, then

$$R = \cosh\left(\frac{\phi}{2}\right) - \hat{B}\sinh\left(\frac{\phi}{2}\right).$$

This representation encapsulates the hyperbolic rotation in the t–x plane, where the rapidity ϕ is related to the velocity v through the relation $\tanh\phi = v/c$. The bivector encapsulates both the plane of the boost and the magnitude of the transformation, thereby enabling a coordinate-free representation of Lorentz boosts.

Computational Implementation of Lorentz Boosts in GA

The computational approach mirrors the algebraic formulation: a Lorentz boost is implemented by constructing a rotor from the exponential map and then applying it to a vector via a sandwich product. In a numerical algorithm the multivectors are typically stored as dictionaries mapping basis blades to their scalar coefficients. Operations such as the geometric product and reversion are implemented algorithmically with careful attention to the underlying signature of the space.

A single function that encapsulates the process of applying a Lorentz boost demonstrates both clarity and rigor. The function computes the rotor corresponding to a boost bivector using hyperbolic functions and applies the sandwich product to obtain the transformed vector. This procedure requires auxiliary routines for the geometric product and reversion, as well as a routine to compute the magnitude of a bivector in accordance with the Minkowski signature.

```
def lorentz_boost(vector, boost_bivector):
    """
    Apply a Lorentz boost to a vector using a boost bivector.

    This function computes a rotor R via the exponential map:
        R = exp(-boost_bivector/2)
    For a boost bivector B, with magnitude phi, the rotor is given
    by:
        R = cosh(phi/2) - (B/phi) * sinh(phi/2)
    The boosted vector is then obtained by evaluating the sandwich
    product:
        v' = R * vector * reverse(R)
    It is assumed that helper functions 'multivector_multiply',
    'reverse', and
    'compute_magnitude' are defined within the computational
    framework.

    Parameters:
        vector (dict): A multivector representing the vector to be
        transformed.
        boost_bivector (dict): A bivector encoding the boost (with
        appropriate Minkowski signature).

    Returns:
        dict: The Lorentz-boosted vector.
    """
    import math
    phi = compute_magnitude(boost_bivector)
    half_phi = phi / 2.0

    # Construct the rotor using hyperbolic functions.
    rotor = {(): math.cosh(half_phi)}
    factor = -math.sinh(half_phi) / (phi if phi != 0 else 1)
    for blade, coeff in boost_bivector.items():
        rotor[blade] = rotor.get(blade, 0) + coeff * factor

    # Apply the sandwich product to obtain the transformed vector.
    temp = multivector_multiply(rotor, vector)
    boosted_vector = multivector_multiply(temp, reverse(rotor))
    return boosted_vector
```

The function above illustrates the encapsulated procedure for computing a Lorentz boost in geometric algebra. The exponential map yields a rotor with hyperbolic cosine and sine components, and the subsequent sandwich product ensures that the transformed vector adheres to the Lorentz invariance of the Minkowski metric. The modular design of the function facilitates integration into a larger computational framework dedicated to relativistic physics simulations.

Python Code Snippet

```python
import math

# Define Minkowski metric for a 4D spacetime (signature: +, -, -, -)
# Here, basis index 0 corresponds to the time-like direction.
metric = {0: 1, 1: -1, 2: -1, 3: -1}

def multiply_blades(b1, b2):
    """
    Multiply two basis blades b1 and b2.
    Each blade is represented as a sorted tuple of basis indices.
    The function returns a tuple (result_blade, factor) where
      result_blade is the
    combined blade (also as a sorted tuple) and factor is the sign
      and metric
    factor arising from reordering and square-of-basis elements.
    """
    result = list(b1)
    factor = 1
    for b in b2:
        # Count the number of elements in 'result' that are greater
        #   than b.
        swaps = sum(1 for x in result if x > b)
        factor *= (-1) ** swaps
        if b in result:
            # If the basis vector already appears, remove it and
            #   introduce the metric factor.
            result.remove(b)
            factor *= metric[b]
        else:
            # Insert b to keep the blade in sorted order.
            inserted = False
            for i, x in enumerate(result):
                if b < x:
                    result.insert(i, b)
                    inserted = True
                    break
            if not inserted:
                result.append(b)
    return tuple(result), factor

def multivector_multiply(mv1, mv2):
    """
    Multiply two multivectors mv1 and mv2.
    Both multivectors are represented as dictionaries mapping a
      basis blade (tuple)
    to its scalar coefficient.
    The function returns a new multivector (dictionary) representing
      their geometric product.
    """
    result = {}
```

```python
        for blade1, coeff1 in mv1.items():
            for blade2, coeff2 in mv2.items():
                new_blade, f = multiply_blades(blade1, blade2)
                result[new_blade] = result.get(new_blade, 0) + coeff1 *
                ↪    coeff2 * f
    return result

def reverse(mv):
    """
    Compute the reversion of a multivector.
    Reversion reverses the order of basis vectors in each blade.
    For a blade of grade r, the reversion introduces a factor
    ↪   (-1)^(r(r-1)/2).
    """
    result = {}
    for blade, coeff in mv.items():
        grade = len(blade)
        rev_factor = (-1) ** (grade * (grade - 1) // 2)
        result[blade] = coeff * rev_factor
    return result

def compute_magnitude(bivector):
    """
    Compute the magnitude of a bivector.
    This is done by computing the geometric product of the bivector
    ↪   with its reverse,
    and taking the square root of the absolute value of the scalar
    ↪   part.
    It is assumed that for a boost bivector the scalar part is
    ↪   positive.
    """
    prod = multivector_multiply(bivector, reverse(bivector))
    scalar = prod.get((), 0)
    return math.sqrt(abs(scalar)) if scalar != 0 else 0

def lorentz_boost(vector, boost_bivector):
    """
    Apply a Lorentz boost to a vector using a boost bivector.

    The boost is implemented via the exponential map:
      R = exp(-boost_bivector/2)
        = cosh(phi/2) - (boost_bivector/phi) * sinh(phi/2),
    where phi is the magnitude of the bivector.
    The transformed vector is obtained by the sandwich product:
      v' = R * vector * reverse(R)

    Parameters:
      vector (dict): The multivector representing the vector (e.g.,
      ↪   a 1-vector).
      boost_bivector (dict): The bivector representing the boost
      ↪   (e.g., in the t-x plane).

    Returns:
```

```python
        dict: The Lorentz-boosted vector.
    """
    phi = compute_magnitude(boost_bivector)
    half_phi = phi / 2.0
    # Construct the rotor using hyperbolic functions.
    rotor = {(): math.cosh(half_phi)}
    factor = -math.sinh(half_phi) / (phi if phi != 0 else 1)
    for blade, coeff in boost_bivector.items():
        rotor[blade] = rotor.get(blade, 0) + coeff * factor

    # Apply the sandwich product: v' = R * vector * reverse(R)
    temp = multivector_multiply(rotor, vector)
    boosted_vector = multivector_multiply(temp, reverse(rotor))
    return boosted_vector

# Example usage:
if __name__ == "__main__":
    # Define a vector in Minkowski space:
    # Represent gamma_0 (time basis vector) as (0,)
    # Represent gamma_1 (x basis vector) as (1,)
    vector = {
        (0,): 1.0,   # time component
        (1,): 0.5    # x component
    }

    # Define a boost bivector in the t-x plane:
    # Represented by the blade (0,1) with a given magnitude
    ↪    (rapidity factor).
    boost_bivector = {
        (0,1): 0.5
    }

    print("Original vector:", vector)
    boosted_vector = lorentz_boost(vector, boost_bivector)
    print("Boosted vector:", boosted_vector)
```

Chapter 21

Complex Structures in GA: Embedding and Computation

Embedding of Complex Numbers in Geometric Algebra

Within the framework of geometric algebra, complex numbers are not treated as an ad hoc extension of the reals but rather emerge naturally through the algebraic structure. A typical complex number can be expressed as

$$z = a + bI,$$

where a and b are real scalars and I represents a unit bivector with the property

$$I^2 = -1.$$

In a two-dimensional Euclidean subalgebra, the bivector I, often identified with the pseudoscalar of that plane, assumes the role of the imaginary unit. The embedding is achieved by interpreting I as an oriented area element rather than an abstract quantity. This embedding preserves the inherent geometric meaning while endowing the complex-like object with clear computational benefits. The coordinate-free representation of rotations in the plane and the simplification of many physical transformations become transparent under this formulation.

Computational Representation and Efficiency

The representation of complex numbers within geometric algebra allows for a unified treatment of rotations and oscillatory phenomena. In computational implementations, complex numbers embedded in GA retain both algebraic consistency and numerical stability. Each complex number is encoded as the sum of a scalar and a bivector, and operations such as multiplication exploit the bivector square property, $I^2 = -1$, to recover the familiar formula

$$(a + bI)(c + dI) = (ac - bd) + (ad + bc)I.$$

This structural similarity not only streamlines algorithmic design but also enables more direct mapping between mathematical theory and numerical simulation. In applications ranging from quantum mechanics to signal processing, the ability to seamlessly integrate complex structures into a broader geometric context translates into reductions in computational overhead and increased clarity in algorithm development.

A function that performs the multiplication of two complex numbers represented in geometric algebra is provided below. This function encapsulates the arithmetic of embedded complex numbers, ensuring that the bivector component consistently behaves as the imaginary unit.

```
def ga_complex_multiply(z1, z2):
    """
    Multiply two complex numbers represented in geometric algebra.

    Each complex number is represented as a dictionary with keys
     'real'
    and 'imag' corresponding to the scalar and bivector parts,
     respectively.
    The bivector functions as the imaginary unit, satisfying I**2 =
     -1.

    For complex numbers z1 = a + b*I and z2 = c + d*I, the product
     is computed as:
        z1 * z2 = (a*c - b*d) + (a*d + b*c)*I

    Parameters:
        z1 (dict): Dictionary representing the first complex number.
        z2 (dict): Dictionary representing the second complex
         number.
```

```
Returns:
    dict: A dictionary with keys 'real' and 'imag' representing
    ↪ the product.
"""
a, b = z1.get('real', 0), z1.get('imag', 0)
c, d = z2.get('real', 0), z2.get('imag', 0)
return {'real': a * c - b * d, 'imag': a * d + b * c}
```

Applications in Computational Physics and Signal Processing

The incorporation of complex structures into geometric algebra is especially advantageous in areas that require a robust representation of phase and rotational symmetry. In computational physics, the encoding of complex amplitudes as a combination of scalar and bivector components facilitates the analysis of wave functions and quantum states. The geometric interpretation of the imaginary unit as an oriented plane element enriches the visualization of phase rotations and cyclic dynamics in multidimensional systems.

In the context of signal processing, the GA embedding permits a unified treatment of Fourier transforms and spectral decompositions. By operating directly on the algebraic elements, algorithms benefit from the intrinsic geometric meaning attached to the complex structure. This promotes cleaner implementations, where the underlying operations such as rotations and reflections are expressed in a coordinate-free and dimension-agnostic manner. The seamless integration of algebra and geometry leads to efficient computations and improved numerical stability in handling oscillatory data.

The explicit representation of complex numbers through geometric algebra thus underpins a range of computational methods, offering both theoretical insight and practical advantages in simulation and analysis.

Python Code Snippet

```
import math

def ga_complex_multiply(z1, z2):
    """
    Multiply two GA-complex numbers.
```

```
    Each complex number is represented as a dictionary with keys
    ↪ 'real'
    and 'imag', where 'real' is the scalar part and 'imag' is the
    ↪ bivector part.
    For z1 = a + b*I and z2 = c + d*I, with I^2 = -1:
        z1 * z2 = (a*c - b*d) + (a*d + b*c)*I
    """
    a, b = z1.get('real', 0), z1.get('imag', 0)
    c, d = z2.get('real', 0), z2.get('imag', 0)
    return {'real': a * c - b * d, 'imag': a * d + b * c}

def ga_complex_add(z1, z2):
    """
    Add two GA-complex numbers.

    Returns:
        (a + c) + (b + d)*I for z1 = a + b*I and z2 = c + d*I.
    """
    return {
        'real': z1.get('real', 0) + z2.get('real', 0),
        'imag': z1.get('imag', 0) + z2.get('imag', 0)
    }

def ga_complex_subtract(z1, z2):
    """
    Subtract z2 from z1.

    Returns:
        (a - c) + (b - d)*I for z1 = a + b*I and z2 = c + d*I.
    """
    return {
        'real': z1.get('real', 0) - z2.get('real', 0),
        'imag': z1.get('imag', 0) - z2.get('imag', 0)
    }

def ga_complex_conjugate(z):
    """
    Compute the conjugate of a GA-complex number.

    For z = a + b*I, the conjugate is:
        z* = a - b*I
    """
    return {'real': z.get('real', 0), 'imag': -z.get('imag', 0)}

def ga_complex_divide(z1, z2):
    """
    Divide two GA-complex numbers.

    The division is defined as:
        z1 / z2 = z1 * conjugate(z2) / |z2|^2
    with:
        |z2|^2 = (c)^2 + (d)^2, for z2 = c + d*I.
```

155

```
    Raises:
        ZeroDivisionError: when the magnitude of z2 is zero.
    """
    conj_z2 = ga_complex_conjugate(z2)
    numerator = ga_complex_multiply(z1, conj_z2)
    denom = z2.get('real', 0)**2 + z2.get('imag', 0)**2
    if denom == 0:
        raise ZeroDivisionError("Division by zero complex number.")
    return {
        'real': numerator.get('real', 0) / denom,
        'imag': numerator.get('imag', 0) / denom
    }

def ga_complex_abs(z):
    """
    Compute the magnitude (absolute value) of a GA-complex number.

    For z = a + b*I, the magnitude is:
        |z| = sqrt(a^2 + b^2)
    """
    return math.sqrt(z.get('real', 0)**2 + z.get('imag', 0)**2)

def ga_complex_exp(z):
    """
    Compute the exponential function for a GA-complex number.

    For z = a + b*I, using Euler's formula:
        exp(z) = exp(a) * (cos(b) + I*sin(b))
    """
    a = z.get('real', 0)
    b = z.get('imag', 0)
    exp_a = math.exp(a)
    return {
        'real': exp_a * math.cos(b),
        'imag': exp_a * math.sin(b)
    }

# Demonstration of the GA-complex arithmetic functions
if __name__ == "__main__":
    # Two example GA-complex numbers: z1 = 2 + 3*I, z2 = 1 - 4*I
    z1 = {'real': 2, 'imag': 3}
    z2 = {'real': 1, 'imag': -4}

    print("z1 =", z1)
    print("z2 =", z2)

    # Multiplication
    product = ga_complex_multiply(z1, z2)
    print("z1 * z2 =", product)

    # Addition
    sum_z = ga_complex_add(z1, z2)
```

```python
print("z1 + z2 =", sum_z)

# Subtraction
difference = ga_complex_subtract(z1, z2)
print("z1 - z2 =", difference)

# Division
division = ga_complex_divide(z1, z2)
print("z1 / z2 =", division)

# Magnitude of z1
magnitude_z1 = ga_complex_abs(z1)
print("|z1| =", magnitude_z1)

# Exponential of z1
exp_z1 = ga_complex_exp(z1)
print("exp(z1) =", exp_z1)
```

Chapter 22

Matrix Representations: Linking GA and Linear Algebra

Matrix Representations of GA Elements

Geometric algebra (GA) elements, including scalars, vectors, and higher-grade multivectors, admit a systematic mapping onto matrix representations. In a well-defined basis, the relationships among various GA elements are preserved by matrices whose entries encode the geometric product. In this formulation, the abstract algebraic structure is translated into the language of linear operators, thereby enabling the application of mature numerical techniques. In particular, the multiplication rules of multivectors are reflected in the arithmetic of matrices, which permits a direct correspondence between the geometric product and matrix multiplication. For instance, a rotor that performs rotations in a plane can be associated with an orthogonal matrix, linking GA operations with conventional rotation matrices.

Matrix Operators and GA Multiplications

The algebraic operations of GA, such as the geometric product, inner product, and outer product, find analogues in the realm of matrix operators. Bivectors, which represent oriented plane segments, can be mapped to skew-symmetric matrices. This identification makes manifest the connection between the commutator of GA elements and the Lie bracket, a familiar concept in linear algebra and Lie theory. Such a mapping bridges abstract algebra with computational methods by allowing multivector products to be recast as products of matrices. This procedure preserves the key properties of GA operations, including associativity and distributivity, and facilitates their numerical evaluation using high-performance linear algebra libraries.

The translation from GA operations to matrices is not merely a mathematical curiosity but is central to many numerical simulations. When GA is applied in physics, the corresponding matrix representations enable the straightforward implementation of transformations and rotations. The structural isomorphism between GA and matrix algebra supports both theoretical insights and the design of optimized algorithms for computational physics and computer science alike.

Algorithmic Implementation of GA Matrix Mappings

A concrete example of bridging GA with matrix representations is provided by the computation of a two-dimensional rotation matrix from a GA rotor. A rotor in two dimensions is typically represented as

$$R = \cos\left(\frac{\theta}{2}\right) - \sin\left(\frac{\theta}{2}\right) I,$$

where I denotes the unit bivector with the property $I^2 = -1$. The corresponding rotation matrix for an angle θ is

$$\begin{pmatrix} \cos\theta & -\sin\theta \\ \sin\theta & \cos\theta \end{pmatrix}.$$

The following function extracts the necessary components from the rotor, computes the full rotation angle, and returns the associated 2×2 matrix.

```
def rotor_to_matrix(rotor):
    """
    Compute the 2D rotation matrix from a GA rotor representation.

    The rotor is represented as a dictionary with keys 'scalar' and
    ↪ 'bivector',
    corresponding to the components of R = cos(theta/2) -
    ↪ sin(theta/2)*I.
    The function returns the 2x2 rotation matrix:

        [ cos(theta)  -sin(theta) ]
        [ sin(theta)   cos(theta) ]

    The angle theta is derived via: theta = 2 * arctan2(s, c), where
    c = rotor['scalar'] and s = -rotor['bivector'] (the sign
    ↪ inversion extracts the
    half-angle sine component).

    Parameters:
        rotor (dict): A dictionary with keys 'scalar' and
        ↪ 'bivector'.

    Returns:
        list: A nested list representing the 2x2 rotation matrix.
    """
    import math
    c = rotor.get('scalar', 1)
    s = -rotor.get('bivector', 0)
    theta = 2 * math.atan2(s, c)
    cos_theta = math.cos(theta)
    sin_theta = math.sin(theta)
    return [[cos_theta, -sin_theta],
            [sin_theta,  cos_theta]]
```

This function demonstrates the extraction of geometric information from a GA rotor and its translation into a conventional matrix form. The use of the arctan2 function ensures correct determination of the rotation angle, thereby preserving the intrinsic geometric meaning of the rotor. The resultant rotation matrix can be subsequently employed in numerical simulations where standard linear algebra techniques are applied.

Matrix representations serve as a robust bridge between the abstract formulation of GA and concrete numerical methods. By utilizing these mappings, operations defined in GA can be computed efficiently with well-established matrix routines, thus facilitating the simulation of complex physical systems while retaining a clear geometric interpretation.

Python Code Snippet

```
import math

def rotor_to_matrix(rotor):
    """
    Compute the 2D rotation matrix from a GA rotor representation.

    The rotor is represented as a dictionary with keys 'scalar' and
       'bivector',
    corresponding to the components of the rotor:

        R = cos(theta/2) - sin(theta/2)*I,

    where I is the unit bivector (with I^2 = -1). The function
       returns the
    2x2 rotation matrix:

        [ cos(theta)  -sin(theta) ]
        [ sin(theta)   cos(theta) ]

    The rotation angle theta is computed from the half-angle
       components via:
        theta = 2 * arctan2(-rotor['bivector'], rotor['scalar'])
    (the minus sign on the bivector extracts the proper half-angle
       sine component).

    Parameters:
        rotor (dict): A dictionary with keys 'scalar' and
           'bivector'.

    Returns:
        list of lists: A 2x2 matrix representing the rotation.
    """
    c = rotor.get('scalar', 1)
    s = -rotor.get('bivector', 0)   # Sign inversion extracts the
       half-angle sine
    theta = 2 * math.atan2(s, c)
    cos_theta = math.cos(theta)
    sin_theta = math.sin(theta)
    return [
        [cos_theta, -sin_theta],
        [sin_theta,  cos_theta]
    ]

def matrix_multiply(A, B):
    """
    Multiply two 2x2 matrices A and B.

    Parameters:
        A (list of lists): The first 2x2 matrix.
        B (list of lists): The second 2x2 matrix.
```

```
    Returns:
        list of lists: The resulting 2x2 matrix product.
    """
    result = [[0, 0], [0, 0]]
    for i in range(2):
        for j in range(2):
            for k in range(2):
                result[i][j] += A[i][k] * B[k][j]
    return result

def matrix_print(M):
    """
    Nicely print a 2x2 matrix M.

    Parameters:
        M (list of lists): The matrix to print.
    """
    for row in M:
        print("[ " + "  ".join(f"{num: .4f}" for num in row) + " ]")

def create_rotor(theta):
    """
    Create a rotor representation from a given rotation angle theta.

    In geometric algebra, a 2D rotor is defined as:
        R = cos(theta/2) - sin(theta/2)*I,
    where I is the unit bivector. Here, we represent the rotor as a
    ↪ dictionary:
        {'scalar': cos(theta/2), 'bivector': sin(theta/2)}

    Note: The sign of 'bivector' is handled in rotor_to_matrix.

    Parameters:
        theta (float): The rotation angle in radians.

    Returns:
        dict: The rotor representation.
    """
    half_angle = theta / 2
    return {
        'scalar': math.cos(half_angle),
        'bivector': math.sin(half_angle)
    }

def main():
    # Example 1: Compute the rotation matrix for a rotation of 45
    ↪ degrees (pi/4 radians)
    theta = math.pi / 4
    rotor = create_rotor(theta)
    print("Rotor representation (GA format) for 45° rotation:")
    print(rotor)
```

```python
    rotation_matrix = rotor_to_matrix(rotor)
    print("\nRotation Matrix derived from the rotor:")
    matrix_print(rotation_matrix)

    # Example 2: Demonstrate matrix multiplication corresponding to
    #   successive rotations.
    # Define two rotations: 30° and 45°.
    theta1 = math.pi / 6  # 30 degrees in radians
    theta2 = math.pi / 4  # 45 degrees in radians
    rotor1 = create_rotor(theta1)
    rotor2 = create_rotor(theta2)

    matrix1 = rotor_to_matrix(rotor1)
    matrix2 = rotor_to_matrix(rotor2)

    # Successive application: first rotate by theta1 and then theta2
    #   corresponds to matrix multiplication:
    product_matrix = matrix_multiply(matrix2, matrix1)

    print("\nRotation Matrix for a 30° rotation:")
    matrix_print(matrix1)

    print("\nRotation Matrix for a 45° rotation:")
    matrix_print(matrix2)

    print("\nCombined Rotation Matrix (applying 30° then 45°):")
    matrix_print(product_matrix)

    # Verify that the combined rotation equals a rotation of (30 +
    #   45) = 75°.
    combined_theta = theta1 + theta2
    expected_matrix = rotor_to_matrix(create_rotor(combined_theta))
    print("\nExpected Rotation Matrix for 75° rotation:")
    matrix_print(expected_matrix)

if __name__ == "__main__":
    main()
```

Chapter 23

Algorithms for the Geometric Product: Implementation Strategies

Mathematical and Computational Foundations

The geometric product, denoted by AB for multivectors A and B, simultaneously encodes both symmetric and antisymmetric components. In practice, this product unifies the inner and outer products under one algebraic operation. When a multivector is expanded in a basis, it takes the form

$$M = \sum_k m_k\, \mathbf{e}_{i_1} \mathbf{e}_{i_2} \cdots \mathbf{e}_{i_k},$$

where $\{\mathbf{e}_i\}$ is an orthonormal basis and each $\mathbf{e}_{i_1} \mathbf{e}_{i_2} \cdots \mathbf{e}_{i_k}$ represents a basis blade. For numerical applications, the blades are converted into computational objects via bit-level encoding; this maps each blade to an integer whose binary representation signifies the presence or absence of a particular $\mathbf{e}_i$. Such representations permit the use of bitwise operations to simulate the algebraic rules of the geometric product with minimal computational overhead.

Bitwise Representations and Table-Driven Approaches

Within this computational framework, the product of two basis blades is determined by applying the bitwise exclusive OR (XOR) operation on their integer representations. The result of this XOR operation uniquely identifies the basis blade corresponding to the product. An essential aspect of the algorithm is the determination of the sign factor resulting from swapping basis vectors to restore canonical order. This sign factor is computed by counting the number of swaps required, wherein each swap introduces a multiplier of -1. Table-driven methods and bitwise approaches are employed to precompute and cache these factors, thus optimizing the overall evaluation of the geometric product.

1 Bitwise Computation of the Sign Factor

The sign correction, which preserves the antisymmetric nature of basis vector multiplication, is determined by examining the bitmask representations of the basis blades. The following Python function iterates over the set bits of one blade and counts the overlaps with lower-order bits in the second blade, thereby computing the parity of swaps required:

```
def compute_sign(blade1, blade2):
    """
    Compute the sign of the geometric product for two basis blades
    ↪ represented as bitmasks.

    The sign is determined by counting the number of basis swaps
    ↪ required to bring the
    product of the blades into canonical order. Each encountered
    ↪ swap contributes a factor
    of -1 to the overall sign. The blade representations are assumed
    ↪ to be integers, where
    each bit corresponds to the presence of a specific basis vector.

    Parameters:
        blade1 (int): Bitmask representing the first basis blade.
        blade2 (int): Bitmask representing the second basis blade.

    Returns:
        int: The sign factor, either +1 or -1.
    """
    sign = 1
    temp = blade1
```

```
while temp:
    lowest = temp & -temp   # Isolate the lowest set bit.
    temp -= lowest
    if bin(blade2 & (lowest - 1)).count("1") % 2:
        sign = -sign
return sign
```

Data Structures and Algorithmic Optimizations in Geometric Algebra

When implementing the geometric product for multivectors, the choice of data structure plays a pivotal role in computational performance. A common strategy is to represent a multivector as a dictionary where each key is an integer encoding a basis blade (via its bitmask) and each corresponding value is the scalar coefficient. This sparse representation enables rapid lookups and minimizes storage in high-dimensional algebras.

Algorithmic optimizations are further achieved through memoization and table-driven techniques. By precomputing multiplication outcomes for standard basis blades and caching intermediate results, redundancies in the computation are reduced significantly. Exploiting parallelism also becomes feasible when independent products of basis blades are computed concurrently, which is particularly advantageous in large-scale numerical simulations.

1 Efficient Implementation of the Geometric Product for Multivectors

For multivectors represented as dictionaries, the geometric product is computed by iterating over all pairs of basis blades from the two factors. The bitwise XOR operation is used to combine the blades, while the sign correction is applied via the function defined previously. The following Python function encapsulates this algorithmic procedure:

```
def geometric_product(mv1, mv2):
    """
    Compute the geometric product of two multivectors represented as
    ↪ dictionaries.

    Each multivector is a mapping from an integer key, which encodes
    ↪ a basis blade using
```

a bitmask representation, to a scalar coefficient. The geometric product is computed
   by iterating over all pairs of basis blades from the two multivectors. The resulting blade
   is determined by the bitwise XOR of the keys, and the associated coefficient is updated
   using the product of the corresponding coefficients and the sign factor computed from the
   basis reordering.

   Parameters:
       mv1 (dict): The first multivector, with keys as integer bitmasks and values as coefficients.
       mv2 (dict): The second multivector, similarly represented.

   Returns:
       dict: A multivector representing the geometric product, stored as a dictionary.
   """
   product = {}
   for blade1, coef1 in mv1.items():
       for blade2, coef2 in mv2.items():
           res_blade = blade1 ^ blade2
           sign = compute_sign(blade1, blade2)
           product[res_blade] = product.get(res_blade, 0.0) + coef1 * coef2 * sign
   return product
```

The amalgamation of bitwise techniques, carefully chosen data structures, and memoization ensures that the implementation of the geometric product meets the rigorous demands of computational physics. The strategies presented here effectively bridge abstract algebraic formulations with high-performance numerical algorithms.

# Python Code Snippet

```
Geometric Algebra Computation: Geometric Product Implementation
This code provides an implementation for calculating the geometric product
of multivectors using bitwise representations.
#
A multivector is represented as a dictionary where each key is an integer bitmask
corresponding to a basis blade, and each value is the scalar coefficient.
#
For example:
```
```

```
#     - The scalar part is represented with key 0 (no bits set).
#     - A basis vector e0 is represented by 1 << 0 (i.e., 1).
#     - A basis vector e1 is represented by 1 << 1 (i.e., 2).
#
# The geometric product is computed by combining blades using the
 ↪    bitwise XOR operation,
# while a sign factor is computed to account for the swaps needed to
 ↪    restore the canonical order.

def compute_sign(blade1, blade2):
    """
    Compute the sign of the geometric product for two basis blades
     ↪    represented as bitmasks.

    The sign is determined by counting the number of basis swaps
     ↪    required to arrange the
    product of the blades into canonical order. Each swap introduces
     ↪    a factor of -1.

    Parameters:
        blade1 (int): Bitmask representing the first basis blade.
        blade2 (int): Bitmask representing the second basis blade.

    Returns:
        int: The sign factor, either +1 or -1.
    """
    sign = 1
    temp = blade1
    while temp:
        lowest = temp & -temp    # Isolate the lowest set bit.
        temp -= lowest
        # Count the number of set bits in blade2 that are lower than
         ↪    the current bit (lowest-1 mask)
        if bin(blade2 & (lowest - 1)).count("1") % 2:
            sign = -sign
    return sign

def geometric_product(mv1, mv2):
    """
    Compute the geometric product of two multivectors represented as
     ↪    dictionaries.

    Each multivector is a mapping from an integer key, which encodes
     ↪    a basis blade via its bitmask,
    to a scalar coefficient. The geometric product is computed by
     ↪    iterating over all pairs of basis
    blades from the two multivectors. The resulting basis blade is
     ↪    obtained by the bitwise XOR of the
    two keys, and the associated coefficient is the product of the
     ↪    original coefficients multiplied
    by a sign factor determined by the basis ordering.

    Parameters:
```

        mv1 (dict): First multivector with keys as integer bitmasks
    ↪   and values as coefficients.
        mv2 (dict): Second multivector, represented similarly.

    Returns:
        dict: A multivector representing the geometric product.
    """
    product = {}
    for blade1, coef1 in mv1.items():
        for blade2, coef2 in mv2.items():
            res_blade = blade1 ^ blade2
            sign = compute_sign(blade1, blade2)
            product[res_blade] = product.get(res_blade, 0.0) + coef1
    ↪   * coef2 * sign
    return product

def format_blade(blade):
    """
    Format a basis blade bitmask into a human-readable string.

    A basis blade is represented as a product of basis vectors. For
    ↪   example, if a blade has the
    binary representation '101', it corresponds to the product e0e2.
    ↪   The scalar (blade 0) is
    represented as "1".

    Parameters:
        blade (int): Integer bitmask of the basis blade.

    Returns:
        str: A string representation of the basis blade.
    """
    if blade == 0:
        return "1"
    basis = []
    index = 0
    while blade:
        if blade & 1:
            basis.append(f"e{index}")
        index += 1
        blade >>= 1
    return "".join(basis)

def print_multivector(mv):
    """
    Print a multivector in a readable format.

    This function sorts the multivector keys and formats each term
    ↪   as the product of the scalar
    coefficient and the corresponding basis blade.

    Parameters:
        mv (dict): The multivector to print.

```python
    """
    terms = []
    for blade in sorted(mv.keys()):
        coef = mv[blade]
        # Filter out coefficients that are negligible (to avoid
        ↪ floating point noise).
        if abs(coef) > 1e-12:
            basis_str = format_blade(blade)
            # Avoid attaching the basis string for scalars.
            term = f"{coef:.2f}" if basis_str == "1" else
            ↪ f"{coef:.2f}{basis_str}"
            terms.append(term)
    print(" + ".join(terms) if terms else "0")

# Example usage: Define two multivectors and compute their geometric
↪ product.
if __name__ == "__main__":
    # Define multivectors using a dictionary representation.
    # For instance, let mv1 = 1 + 2e0, where:
    #   - 1 is the scalar part, represented by key 0.
    #   - e0 is represented by key 1 << 0, which is 1.
    mv1 = {
        0: 1.0,   # Scalar part: 1.0
        1: 2.0    # e0: 2.0e0
    }

    # Let mv2 = 3 + 4e1, where:
    #   - 3 is the scalar part, represented by key 0.
    #   - e1 is represented by key 1 << 1, which is 2.
    mv2 = {
        0: 3.0,   # Scalar part: 3.0
        2: 4.0    # e1: 4.0e1 (since 1 << 1 = 2)
    }

    # Compute the geometric product of mv1 and mv2.
    gp = geometric_product(mv1, mv2)

    # Display the multivectors and their geometric product.
    print("Multivector mv1:")
    print_multivector(mv1)

    print("\nMultivector mv2:")
    print_multivector(mv2)

    print("\nGeometric Product (mv1 * mv2):")
    print_multivector(gp)
```

Chapter 24

Symbolic Manipulation in GA: Tools and Techniques

Foundations of Symbolic Computation in Geometric Algebra

Geometric algebra provides a unifying framework where multivectors, such as

$$M = \sum_i m_i B_i,$$

are expressed in terms of scalar coefficients m_i and basis blades B_i. The symbolic representation of these entities enables precise analytical derivations and manipulations. Symbolic computation in geometric algebra involves reformulating and simplifying expressions that contain the geometric product, denoted by juxtaposition, which combines both symmetric and antisymmetric components. This procedure usually encompasses the expansion of products, the enforcement of anticommutation relations

$$e_i e_j = -e_j e_i \quad \text{for } i \neq j,$$

and the consolidation of like terms into canonical forms. Such symbolic strategies support exact evaluation of expressions that arise in theoretical derivations and algorithmic developments.

Software Tools and Libraries Supporting Symbolic GA

A number of sophisticated symbolic computation libraries have been developed within the broader computer algebra community. These packages are capable of handling algebraic structures that extend beyond conventional commutative algebras. Notable examples include systems that build on the symbolic manipulation capabilities of libraries, such as Sympy. These tools allow multivector expressions to be defined as symbolic sums, products, and functionals, thereby facilitating operations including expansion, factorization, and canonicalization.

In the context of geometric algebra, the symbolic framework must accommodate the non-commutative nature of the geometric product and reconcile the antisymmetry inherent in the product of basis vectors. By representing basis blades as distinct symbolic objects and assigning appropriate algebraic properties to them, software tools enable automated transformation and reduction of complex expressions, ensuring that every term is positioned in a well-defined order and redundant contributions are eliminated.

Algorithmic Strategies for Symbolic Simplification

Algorithmic simplification in geometric algebra centers on the systematic application of rewriting rules to obtain canonical forms. These strategies include recursive pattern matching, substitution of equivalent expressions, and consolidation of coefficients for equivalent basis blades. A well-designed symbolic system must rigorously enforce the rules of the geometric product: for instance, the anti-commutation rule

$$e_i e_j = -e_j e_i,$$

and the idempotence of basis vectors when squared (under a positive metric), i.e.,

$$e_i^2 = 1.$$

The process integrates both rule-based and table-driven approaches to ensure that each term in a symbolic multivector is represented in a standardized order. The algorithms also account for sign changes induced by the reordering of basis elements, a critical aspect that affects the overall consistency of the computation.

1 Function for Symbolic Simplification

A representative interface in the software integrates the underlying symbolic algebra library with the geometric algebra framework by abstracting the complexity of term manipulation. The following Python function encapsulates the simplification process for a symbolic geometric algebra expression. It is presented in a self-contained snippet that illustrates the intended functionality:

```
def symbolic_simplify(expression):
    """
    Simplify a symbolic geometric algebra expression.

    The function accepts an expression constructed from symbolic
    ↪ multivector terms
    and applies a set of rewriting rules to achieve a canonical
    ↪ form. The process
    leverages an underlying symbolic algebra library to combine like
    ↪ terms, enforce
    the antisymmetry of the geometric product, and eliminate
    ↪ redundant contributions.

    Parameters:
        expression: A symbolic expression representing a geometric
        ↪ algebra multivector.

    Returns:
        A simplified symbolic expression with canonical ordering of
        ↪ basis blades.
    """
    # The implementation uses the simplify method of the symbolic
    ↪ algebra library.
    return expression.simplify()
```

This function exemplifies the abstraction of complex symbolic manipulations into an interface that automatically enforces the algebraic identities and anticommutation properties of geometric algebra. The systematic rewriting of terms ensures that expressions are reduced to their simplest forms, which is essential for both analytical insights and subsequent numerical evaluations.

Python Code Snippet

```
import sympy as sp

def canonical_blade(blade_list):
```

```
"""
Given a list of basis indices (which may be unsorted and include
    repeats),
this function returns the canonical representation of the blade
    as a
sorted tuple with all duplicate pairs canceled (using e_i*e_i =
    1)
and a sign factor resulting from the reordering.
"""
# Make a copy to avoid modifying the original list
L = list(blade_list)
sign = 1
n = len(L)
# Bubble sort to reorder L into increasing order and count
    swaps.
for i in range(n):
    for j in range(0, n - i - 1):
        if L[j] > L[j+1]:
            L[j], L[j+1] = L[j+1], L[j]
            sign *= -1
# Cancel duplicate basis elements (since e_i^2 = 1)
canonical = []
i = 0
while i < len(L):
    if i + 1 < len(L) and L[i] == L[i+1]:
        # Duplicate pair cancels to 1 (with no sign change)
        i += 2
    else:
        canonical.append(L[i])
        i += 1
return (tuple(canonical), sign)

class Multivector:
    """
    A class to represent multivectors in geometric algebra.

    The multivector is stored as a dictionary mapping a canonical
        blade
    (represented by a tuple of basis indices; the empty tuple
        corresponds
    to the scalar part) to its coefficient, which is a sympy
        expression.
    """
    def __init__(self, terms=None):
        # terms is a dictionary: key = blade (tuple), value =
            coefficient (sympy expression).
        if terms is None:
            self.terms = {}
        else:
            self.terms = dict(terms)

    def __add__(self, other):
        result = Multivector(self.terms)
```

```python
        for blade, coeff in other.terms.items():
            if blade in result.terms:
                result.terms[blade] += coeff
            else:
                result.terms[blade] = coeff
        # Simplify coefficients and remove terms with zero
        ↪  coefficients.
        result.terms = {b: sp.simplify(c) for b, c in
        ↪  result.terms.items() if not sp.simplify(c) == 0}
        return result

    def __rmul__(self, other):
        """
        Allow left scalar multiplication.
        """
        if isinstance(other, (sp.Basic, int, float)):
            new_terms = {b: sp.simplify(other * coeff) for b, coeff
            ↪  in self.terms.items()}
            return Multivector(new_terms)
        else:
            return NotImplemented

    def __mul__(self, other):
        """
        Geometric product between two multivectors.
        For each term, the product rule uses the canonical_blade
        ↪  function to
        reorder the concatenated list of basis indices, enforcing
        ↪  e_i*e_j = -e_j*e_i
        for i  j and e_i^2 = 1.
        """
        result = Multivector()
        for blade_a, coeff_a in self.terms.items():
            for blade_b, coeff_b in other.terms.items():
                # Concatenate the two blades as lists of indices.
                combined = list(blade_a + blade_b)
                can_blade, sign = canonical_blade(combined)
                new_coeff = sp.simplify(coeff_a * coeff_b * sign)
                if can_blade in result.terms:
                    result.terms[can_blade] += new_coeff
                else:
                    result.terms[can_blade] = new_coeff
        # Remove any zero coefficients after multiplication.
        result.terms = {b: sp.simplify(c) for b, c in
        ↪  result.terms.items() if not sp.simplify(c)==0}
        return result

    def simplify(self):
        """
        Simplify the multivector by combining like terms and
        ↪  simplifying coefficients.
        """
```

```python
            self.terms = {b: sp.simplify(c) for b, c in
            ↪  self.terms.items() if not sp.simplify(c)==0}
            return self

    def __repr__(self):
        if not self.terms:
            return "0"
        term_strs = []
        # Sort terms by grade and then by the blade tuple.
        for blade, coeff in sorted(self.terms.items(), key=lambda x:
        ↪  (len(x[0]), x[0])):
            if blade == ():
                blade_str = "1"
            else:
                blade_str = "".join("e{}".format(i) for i in blade)
            term_strs.append(f"({coeff}){blade_str}")
        return " + ".join(term_strs)

def symbolic_simplify(expression):
    """
    Simplify a symbolic geometric algebra expression.

    This function accepts an expression constructed as a Multivector
    and applies rewriting rules to enforce a canonical form. It
    ↪  leverages
    the Multivector's own simplify method.

    Parameters:
        expression: A Multivector representing a geometric algebra
        ↪  expression.

    Returns:
        A simplified Multivector with canonical ordering of basis
        ↪  blades.
    """
    return expression.simplify()

# Example usage and demonstration of the symbolic GA operations
if __name__ == "__main__":
    # Define symbolic coefficients.
    a, b, c = sp.symbols('a b c', real=True)

    # Define basis vectors as multivectors.
    # Each basis vector e_i is represented as a multivector with
    ↪  blade (i,).
    e1 = Multivector({(1,): 1})
    e2 = Multivector({(2,): 1})
    e3 = Multivector({(3,): 1})

    # Construct a multivector M = a + b*e1 + c*(e1*e2)
    M = Multivector({(): a}) + b * e1 + c * (e1 * e2)
    print("Multivector M:")
    print(M)
```

```python
# Compute the geometric product of two basis vectors.
# According to the anticommutation rule, e2 * e1 should equal -
↪  (e1 * e2).
N = e2 * e1
print("\nGeometric product of e2 and e1 (N):")
print(N)

# Demonstrate symbolic simplification:
# The sum (e1*e2 + e2*e1) should simplify to 0 due to
↪  antisymmetry.
expr = (e1 * e2) + (e2 * e1)
simplified_expr = symbolic_simplify(expr)
print("\nSimplified expression of (e1*e2 + e2*e1):")
print(simplified_expr)
```

Chapter 25

Basis Transformations: Changing Perspectives in GA

Mathematical Foundations of Basis Transformations

In the framework of geometric algebra, a basis transformation is construed as a change from one set of generating vectors to another. Consider two sets of basis vectors, $\{e_i\}$ and $\{f_j\}$, with the transformation defined by

$$f_j = \sum_i T^i_j e_i,$$

where T^i_j are the components of the transformation matrix. Such a shift of basis is not only a matter of coordinate reparameterization but also a process that preserves the algebraic structure implicit in the geometric product and the graded decomposition of multivectors. The transformation induces a new representation of all basis blades, and therefore, any multivector

$$M = \sum_\alpha m_\alpha B_\alpha,$$

must be re-expressed in terms of the new blades. The inherent antisymmetry present in the wedge product and the sorted structure of blades imply that every permutation of basis vectors under the transformation must be accompanied by appropriate sign adjustments.

Representation of Multivectors Under Basis Change

A multivector, when expressed in an original basis, is decomposed into a linear combination of blades. Under a basis transformation, each blade B_α transforms in accordance with the multilinear extension of the mapping of individual basis vectors. For instance, a bivector constructed as

$$B = e_i \wedge e_j,$$

transforms to

$$B' = f_i \wedge f_j = \left(\sum_k T_i^k e_k\right) \wedge \left(\sum_l T_j^l e_l\right),$$

where the determinant of the transformation restricted to the indices involved, along with the reordering of the constituent vectors, governs the sign and magnitude of the new blade. The computational evaluation of such expressions requires careful management of the degree, sign due to anticommutation, and the algebraic combination of resulting coefficients.

Computational Strategies for Basis Transformations

The process of converting a multivector from one basis to another entails a systematic algorithm that applies the linear transformation to the individual basis vectors and then reconstructs the multivector by recomputing the geometric product in the new basis. This operation leverages the fact that the geometric product is linear and that basis blades obey a canonical ordering, which is essential for ensuring a unique representation.

A central challenge lies in handling the reordering of basis elements, since the antisymmetry of the wedge product demands that each swap of indices introduces a sign change. An effective algorithm will traverse the set of basis indices for every blade, substitute each vector according to the transformation matrix, and then reassemble the product while consolidating like terms. This procedure not only maintains the mathematical rigor of the transformation but also optimizes symbolic and numerical computations by reordering the terms into a canonical form.

Implementation Example: A Basis Transformation Function

An example of a computational strategy for basis transformation is encapsulated in a Python function. This function accepts a multivector, represented as a dictionary where keys denote ordered tuples corresponding to basis blades and values represent coefficients, along with a transformation matrix T. The transformation matrix maps the original basis to a new set of basis vectors via the relation

$$f_j = \sum_i T[i][j]\, e_i.$$

The function then recomputes the representation of the multivector in the new basis by applying the transformation to each basis vector and adjusting the resulting blade with the appropriate sign factors. A detailed explanation is embedded in the function's documentation string.

```
def transform_basis(multivector, T):
    """
    Transform a multivector to a new basis using the transformation
    ↪ matrix T.

    This function accepts a multivector expressed in an initial
    ↪ basis and
    applies a linear transformation to each of its basis vectors.
    ↪ The transformation
    is defined by the relation f_j = sum_i T[i][j] * e_i, where T is
    ↪ the transformation
    matrix mapping the original basis to the new basis. For each
    ↪ blade in the multivector,
    the function computes the new representation, taking into
    ↪ account the necessary
    reordering of vectors and the sign adjustments resulting from
    ↪ the antisymmetry of
    the wedge product. The input multivector is assumed to be
    ↪ represented as a dictionary
```

```
        with keys as ordered tuples of basis indices and values as the
    ↪    corresponding coefficients.

    Parameters:
        multivector: dict
            A dictionary representation of a multivector where each
            ↪  key is an ordered tuple
            indicating a basis blade and each value is the
            ↪  associated coefficient.
        T: list of lists or 2D array
            The transformation matrix with T[i][j] providing the
            ↪  coefficient for mapping
            the original basis vector e_i to the new basis vector
            ↪  f_j.

    Returns:
        dict
            A dictionary representing the multivector in the new
            ↪  basis, maintaining the same
            structure as the input.
    """
    new_multivector = {}
    # The transformation algorithm is implemented here. For each
    ↪  basis blade,
    # the function applies the matrix T to each component of the
    ↪  blade, reorders the
    # resulting vectors into canonical form (accounting for sign
    ↪  changes due to anticommutation),
    # and aggregates coefficients corresponding to identical
    ↪  resulting blades.
    return new_multivector
```

The function serves as a prototype for automating basis transformations within a geometric algebra computational framework. Its design underscores the importance of maintaining canonical order, enforcing sign conventions, and reassembling the transformed blades into a coherent multivector representation, all of which are critical for both analytical and numerical evaluations in geometric algebra.

Python Code Snippet

```
# Comprehensive Python Code Snippet for Basis Transformations in
↪  Geometric Algebra

import itertools
import numpy as np
```

```python
def transform_basis(multivector, T):
    """
    Transform a multivector to a new basis using the transformation
    ↪  matrix T.

    In geometric algebra, the transformation between two bases is
    ↪  governed by
    the relation:
        f_j = sum_i T[i][j] * e_i,
    where { e_i } is the original basis and { f_j } is the new
    ↪  basis.

    To express a multivector originally represented in the e-basis
    ↪  in terms of the f-basis,
    we substitute each occurrence of e_i in the multivector by its
    ↪  expression in f:
        e_i = sum_j T_inv[j][i] * f_j,
    where T_inv is the inverse of T.

    For each blade in the multivector (represented as an ordered
    ↪  tuple of indices corresponding
    to the original basis), the function computes the multilinear
    ↪  expansion by applying the
    inverse transformation to each basis vector. Since the wedge
    ↪  product is antisymmetric,
    the new set of basis indices must be reordered into canonical
    ↪  (sorted) form, and an associated
    sign factor must be applied.

    Parameters:
        multivector: dict
            A dictionary representation of a multivector where each
            ↪  key is an ordered tuple
            indicating a basis blade (e.g., (0, 1) representing e0
            ↪  e1) of the original basis,
            and each value is the corresponding coefficient.
        T: list of lists or 2D array
            The transformation matrix such that T[i][j] provides the
            ↪  coefficient mapping the new
            basis vector f_j to the original basis vector e_i
            ↪  according to:
                f_j = sum_i T[i][j] * e_i.

    Returns:
        dict
            A dictionary representing the multivector in the new
            ↪  f-basis, maintaining the same
            structure as the input, with keys as ordered tuples of
            ↪  new basis indices.
    """
    # Convert T to a numpy array and compute its inverse.
    T_np = np.array(T, dtype=float)
    T_inv = np.linalg.inv(T_np)
```

```python
new_multivector = {}

def canonicalize(blade):
    """
    Reorder the blade tuple into its canonical (sorted) form.
    Returns:
        canonical_blade: tuple
            The sorted tuple representing the blade.
        sign: int
            The sign factor (+1 or -1) arising from the
            ↪ necessary permutation.
    """
    sorted_blade = tuple(sorted(blade))
    sign = 1
    blade_list = list(blade)
    # Compute sign of permutation by counting inversions
    for i in range(len(blade_list)):
        for j in range(i + 1, len(blade_list)):
            if blade_list[i] > blade_list[j]:
                sign *= -1
    return sorted_blade, sign

# Iterate over each blade in the original multivector.
for blade, coeff in multivector.items():
    grade = len(blade)
    # For each occurrence of e_i in the blade, substitute:
    #   e_i = sum_j T_inv[j][i] * f_j
    # This results in a multilinear expansion over the new basis
    ↪ indices.
    for new_indices in itertools.product(range(T_np.shape[1]),
    ↪ repeat=grade):
        product_coeff = 1.0
        for r, new_index in enumerate(new_indices):
            product_coeff *= T_inv[new_index, blade[r]]
        # Skip terms with negligible contribution.
        if abs(product_coeff) < 1e-12:
            continue
        # Reorder the resulting tuple into canonical form and
        ↪ adjust the sign.
        new_blade, sign = canonicalize(new_indices)
        total_coeff = coeff * product_coeff * sign
        # Aggregate coefficients for terms corresponding to the
        ↪ same basis blade.
        if new_blade in new_multivector:
            new_multivector[new_blade] += total_coeff
        else:
            new_multivector[new_blade] = total_coeff

# Remove terms with near-zero coefficients for numerical
↪ stability.
new_multivector = {blade: c for blade, c in
↪ new_multivector.items() if abs(c) > 1e-12}
```

```
    return new_multivector

# Example usage:
if __name__ == "__main__":
    # Define a sample multivector in a 3-dimensional space using the
    ↪ original basis e0, e1, e2.
    # The multivector is represented as a dictionary:
    #  - The scalar part is denoted by the empty tuple ().
    #  - Vectors are represented as single-element tuples, e.g.,
    ↪ (0,) for e0.
    #  - Bivectors and higher-grade elements are represented as
    ↪ tuples of indices.
    multivector = {
        (): 3.0,            # Scalar part
        (0,): 1.0,          # Vector component along e0
        (1,): 2.0,          # Vector component along e1
        (0,1): 4.0,         # Bivector part: e0 e1
        (0,1,2): 5.0        # Trivector part (pseudoscalar in 3D)
    }

    # Define a transformation matrix T for the 3D space.
    # The matrix maps new basis vectors f_j to the original basis
    ↪ e_i:
    #       f_j = sum_i T[i][j] * e_i.
    # For example, the given T scales e1 by 2 and e2 by 3, leaving
    ↪ e0 unchanged.
    T = [
        [1.0, 0.0, 0.0],
        [0.0, 2.0, 0.0],
        [0.0, 0.0, 3.0]
    ]

    transformed_multivector = transform_basis(multivector, T)
    print("Transformed multivector in the new basis (f):")
    for blade, coeff in transformed_multivector.items():
        print(f"Blade {blade}: {coeff}")
```

Chapter 26

Duality in Geometric Algebra: Theoretical Foundations

Mathematical Basis of Duality

Within the framework of geometric algebra, duality is defined as an operation that maps a blade of grade k in an n-dimensional space to its complementary blade of grade $n - k$. This relationship is established through contraction with the pseudoscalar, denoted by I, which is the unique element of maximal grade in the algebra. For a given blade B_k, the dual is defined as

$$B_k^* = B_k I^{-1},$$

where I^{-1} exists provided that the pseudoscalar is non-singular. In many formulations, the dual of a multivector M, expressed as

$$M = \sum_{k=0}^{n} \langle M \rangle_k,$$

is computed by applying the duality operation grade by grade. An important algebraic property is that a double application of the dual operation recovers the original element up to a sign factor:

$$\left(B_k^*\right)^* = (-1)^{k(n-k)} B_k.$$

This involutive feature underlies the symmetry inherent in the geometric structure and provides an algebraic tool for interchanging subspaces of complementary dimensions.

Physical Interpretations and Significance

Duality in geometric algebra encapsulates a profound correspondence between geometrically complementary entities. In physical contexts, bivectors that represent oriented planes may be dualized to yield vectors orthogonal to those planes. For instance, when considering electromagnetic theory, the bivector representation of the electromagnetic field can be mapped to a dual form that isolates the electric or magnetic field components in a Lorentz-invariant manner. This interplay is reminiscent of Hodge duality in differential forms and offers an efficient means to encapsulate conservation laws and symmetry operations. The complementary nature of the dual operation provides a natural language for expressing the orthogonality and complementarity of physical observables, thereby facilitating the mathematical description of phenomena ranging from classical field theory to quantum mechanics.

Computational Implementation of Duality in Geometric Algebra

In computational implementations, multivectors are often represented by data structures that encode the coefficient associated with each basis blade. The dual operation is performed by multiplying each element by the inverse of the pseudoscalar and by appropriately reordering the indices of every blade to conform to a canonical representation. Such a procedure ensures that the resulting multivector preserves the algebraic structure and grading dictated by the dual transformation.

A function that computes the dual of a multivector may be encapsulated in a single, well-documented Python function. An example, provided below, illustrates this process. The function assumes that the multivector is stored in a dictionary with keys defined by tuples of indices and values corresponding to coefficients, while the pseudoscalar is represented as a numerical value for normalization purposes.

```
def compute_dual(multivector, I):
    """
    Compute the dual of a multivector using the pseudoscalar I.

    In geometric algebra, the dual operation is defined by:
        M* = M * I^{-1},
    where I is the pseudoscalar of the algebra. The multivector is
    ↪ assumed to
    be represented as a dictionary with keys as tuples representing
    ↪ basis blades
    and values as the corresponding coefficients. The function
    ↪ multiplies each
    coefficient by the inverse of I and reorders the basis indices
    ↪ into canonical
    form, thereby mapping each k-blade into its complementary
    ↪ (n-k)-blade.

    Parameters:
        multivector (dict): The input multivector with keys as
        ↪ tuples (basis blades)
                            and values as coefficients.
        I (float): The pseudoscalar value used for the dual
        ↪ transformation.

    Returns:
        dict: The dual multivector in the same dictionary format.
    """
    dual = {}
    inv_I = 1.0 / I
    for blade, coeff in multivector.items():
        # The canonical form is obtained by sorting the indices.
        dual_blade = tuple(sorted(blade))
        dual[dual_blade] = coeff * inv_I
    return dual
```

The implementation of the dual operation as detailed above captures the essential algebraic transformations while remaining efficient for numerical evaluations in physical simulations and symbolic computations alike.

Python Code Snippet

```
# This comprehensive Python code implements key equations and
↪ algorithms
# related to the duality operation in geometric algebra as described
↪ in this chapter.
#
# In an n-dimensional space, the dual of a homogeneous blade B
↪ (represented by a
```

```
# tuple of basis indices) is computed via:
#
#     B* = (B) * (1 / I) * ()
#
# where B is the complementary blade (i.e. the ordered set of basis
↪  indices
# not present in B with respect to the full index set {1,2,...,n}),
#   is the sign of the permutation required to reorder (B, B) into
↪  the standard
# ordered tuple (1,2,...,n), and I is the pseudoscalar. Furthermore,
↪  the key
# algebraic property of the dual is that a double dual recovers the
↪  original blade
# up to a sign:
#
#     (B*)* = (-1)^(k*(n-k)) * B,
#
# where k is the grade (number of basis vectors) of B.
#
# The following code defines functions to compute the dual, evaluate
↪  the double
# dual (involution), and verify the involutive property for a sample
↪  multivector.

def permutation_sign(seq):
    """
    Compute the sign (+1 or -1) of the permutation required to sort
    ↪  the sequence.
    This function counts the number of inversions in the list.
    """
    sign = 1
    seq = list(seq)
    n = len(seq)
    # Count how many pairs are out of order.
    for i in range(n):
        for j in range(i+1, n):
            if seq[i] > seq[j]:
                sign *= -1
    return sign

def complement_and_sign(blade, n):
    """
    Given a blade represented as a tuple of indices (e.g., (1, 2)),
    compute its complementary blade and the associated permutation
    ↪  sign.

    The complementary blade is the sorted tuple of indices from the
    ↪  full set
    {1, 2, ..., n} that are not in the original blade.

    The permutation sign is computed as the sign of the permutation
    ↪  that reorders
```

```
            the concatenated tuple (blade + complement) to the canonical
         ↪  order (1,2,...,n).

        Returns:
            (complement, sign) where:
                complement (tuple): the complementary blade,
                sign (int): +1 or -1.
        """
        full = tuple(range(1, n+1))
        blade = tuple(sorted(blade))
        # Compute complement: all indices in the full space not present
         ↪ in blade.
        comp = tuple(sorted(set(full) - set(blade)))
        # Concatenate blade and its complement.
        concatenated = list(blade) + list(comp)
        # The sign is the parity of the permutation that sorts
         ↪ 'concatenated' to 'full'.
        sign = permutation_sign(concatenated)
        return comp, sign

    def compute_dual(multivector, I, n):
        """
        Compute the dual of a multivector using the pseudoscalar I in an
         ↪ n-dimensional space.

        For each homogeneous blade B in the multivector, the dual is
         ↪ defined as:
            B* = (B complementary) * (1/I) * (permutation sign)
        The multivector is represented as a dictionary where keys are
         ↪ tuples (basis blades)
        and values are their corresponding coefficients.

        Parameters:
            multivector (dict): Input multivector.
            I (float): Scalar value of the pseudoscalar (assumed
             ↪ non-zero).
            n (int): Dimension of the vector space.

        Returns:
            dict: A new multivector dictionary representing the dual.
        """
        dual = {}
        inv_I = 1.0 / I
        for blade, coeff in multivector.items():
            # Compute the complementary blade and the permutation sign.
            comp, sign = complement_and_sign(blade, n)
            dual_blade = comp   # The dual blade is defined as the
             ↪ complement.
            # Multiply coefficient by the permutation sign and I
             ↪ inverse.
            dual_coeff = coeff * sign * inv_I
            # Accumulate the coefficient if the same dual blade appears
            if dual_blade in dual:
```

```
                dual[dual_blade] += dual_coeff
            else:
                dual[dual_blade] = dual_coeff
    return dual

def compute_double_dual(multivector, I, n):
    """
    Compute the double dual (dual of the dual) of a multivector.

    According to the duality theorem for a homogeneous blade of
    ↪ grade k in an
    n-dimensional space:
        (B*)* = (-1)^(k*(n-k)) * B.

    The function applies the dual operation twice.
    """
    first_dual = compute_dual(multivector, I, n)
    double_dual = compute_dual(first_dual, I, n)
    return double_dual

def grade(blade):
    """
    Compute the grade of a blade.

    The grade is simply the length of the tuple representing the
    ↪ blade.
    """
    return len(blade)

def sign_factor(blade, n):
    """
    Compute the sign factor (-1)^(k*(n-k)) for a blade of grade k
    in an n-dimensional space.
    """
    k = grade(blade)
    return (-1)**(k*(n - k))

def verify_involution(multivector, I, n):
    """
    Verify the double dual involution property for each homogeneous
    ↪ component
    of a multivector.

    Specifically, for every blade B in the multivector it checks:
        (B*)* = (-1)^(k*(n-k)) * B,
    where k is the grade of B.

    Prints verification messages for each blade and returns True if
    ↪ all
    components satisfy the property.
    """
    double_dual = compute_double_dual(multivector, I, n)
    success = True
```

```python
        for blade, coeff in multivector.items():
            factor = sign_factor(blade, n)
            expected = coeff * factor
            # Use canonical (sorted) representation of the blade.
            canonical = tuple(sorted(blade))
            computed = double_dual.get(canonical, 0)
            if abs(computed - expected) > 1e-10:
                print(f"Mismatch for blade {blade}: expected {expected},
                ↪    got {computed}")
                success = False
            else:
                print(f"Blade {blade} verified: expected {expected}, got
                ↪    {computed}")
    if success:
        print("\nInvolution property verified for all blades.")
    else:
        print("\nInvolution property failed for some blades.")
    return success

def main():
    """
    Main routine demonstrating the computation of the dual and
    ↪    double dual
    of a multivector in geometric algebra.

    An example multivector is defined in a 3-dimensional space with
    ↪    a normalized
    pseudoscalar (I = 1.0). The multivector consists of various
    ↪    grade components:
      - Scalar: represented by an empty tuple ()
      - Vectors: e.g., (1,) represents basis vector e1
      - Bivectors: e.g., (1,2) represents e1 ^ e2
      - Pseudoscalar: e.g., (1,2,3) represents e1 ^ e2 ^ e3
    """
    n = 3  # Dimension of the space.
    # For a normalized pseudoscalar, we choose I = 1.0.
    I = 1.0

    # Define a sample multivector as a dictionary.
    multivector = {
        (): 2.0,         # Scalar component.
        (1,): 3.0,       # Vector component e1.
        (2,): -1.5,      # Vector component e2.
        (1,2): 4.0,      # Bivector component e1 ^ e2.
        (1,2,3): -0.5    # Pseudoscalar component e1 ^ e2 ^ e3.
    }

    print("Original multivector:")
    for blade, coeff in multivector.items():
        print(f"Blade {blade}: {coeff}")

    # Compute and display the dual of the multivector.
    dual_mv = compute_dual(multivector, I, n)
```

```python
        print("\nDual multivector:")
        for blade, coeff in dual_mv.items():
            print(f"Blade {blade}: {coeff}")

        # Compute and display the double dual of the multivector.
        double_dual_mv = compute_double_dual(multivector, I, n)
        print("\nDouble dual multivector:")
        for blade, coeff in double_dual_mv.items():
            print(f"Blade {blade}: {coeff}")

        # Verify that the double dual satisfies the involution property.
        print("\nVerifying the double dual involution property:")
        verify_involution(multivector, I, n)

if __name__ == "__main__":
    main()
```

Chapter 27

Projectors: Extracting Subspace Components

Mathematical Formulation of Projection Operators

In geometric algebra, projectors provide a systematic means to extract the component of a multivector that lies within a specified subspace. Given an invertible blade B representing a subspace, the projection of a vector v onto this subspace is mathematically defined as

$$v_\parallel = (v \cdot B) B^{-1},$$

where the inner product $v \cdot B$ contracts v with B, and B^{-1} denotes the multiplicative inverse of the blade. The operator defined by this expression is idempotent, meaning that applying the projection operation twice yields the same result:

$$\left(v_\parallel\right)_\parallel = v_\parallel.$$

Such a property is critical in many physical applications, for instance, when decomposing force vectors into components parallel and perpendicular to a surface, or isolating field contributions along preferred directions. The formalism extends beyond vectors to more general multivectors, allowing for the isolation of components of arbitrary grade with respect to a given subspace.

Algorithmic Considerations in Projection Operations

The computational realization of projection operators necessitates precise manipulation of algebraic entities. The evaluation involves two essential operations: the contraction via the inner product and the computation of the inverse of an invertible blade. In numerical and symbolic implementations, multivectors are typically represented through data structures that map basis blades to coefficients. The projection operation must preserve the algebraic consistency of these representations while ensuring robustness and efficiency.

A representative implementation in Python encapsulates the projection of a vector onto a subspace through a dedicated function. This function adheres to the mathematical definition by first computing the inner product between the vector and the blade, followed by multiplication with the blade's inverse. The following code snippet illustrates this procedure:

```
def project_vector(v, B):
    """
    Compute the projection of a vector v onto the subspace defined
      by an invertible blade B.

    The projection is defined as:
        v_parallel = (v · B) * B^{-1},
    where v is a multivector representing a vector, B is an
      invertible blade,
    and B^{-1} represents the multiplicative inverse of B.

    The function assumes the existence of a 'dot' operator to
      compute the inner product
    between multivectors and an 'inverse' operator to determine the
      inverse of the blade.

    Parameters:
        v (Multivector): The input vector.
        B (Multivector): An invertible blade defining the target
          subspace.

    Returns:
        Multivector: The component of v that lies within the
          subspace represented by B.
    """
    return dot(v, B) * inverse(B)
```

This function encapsulates the primary steps of the projection algorithm, ensuring that the underlying operations conform to the algebraic principles of geometric algebra.

Properties and Applications of Projection Operators

Projectors in geometric algebra are characterized by their idempotent nature, a property that mathematically confirms the consistency of the projection process. For any projection operator P, the relation

$$P^2 = P$$

holds, which guarantees that once a multivector component has been extracted, additional applications of the projection do not further modify it. This characteristic is indispensable in physical simulations where repeated decompositions or iterative methods are employed.

The application of projection operators extends to various domains in physics. In classical mechanics, projectors are used to decompose motion into components that are parallel and perpendicular to constraints. In electromagnetism, they facilitate the separation of electric and magnetic field components in different inertial frames. Moreover, in quantum mechanics, projectors serve to isolate state components corresponding to distinct measurement outcomes. The algorithmic implementation of these projection methods bridges theoretical formulation with computational efficiency, thereby supporting extensive simulation and analysis in both physics and computational science.

By integrating algebraic rigor with efficient computation, the formulation and implementation of projection operators provide a versatile framework for isolating subspace components, a task central to the modeling and simulation of complex physical phenomena.

Python Code Snippet

```
#!/usr/bin/env python3
"""
A self-contained implementation of a minimal 3D geometric algebra
    (Cl(3,0))
to demonstrate the projection operator in geometric algebra:
```

```
    v_parallel = (v · B) * B^{-1}

Here, v is a vector (grade-1 multivector) and B is an invertible
↪ blade
(e.g. a bivector representing a plane). The inner product '·' is
↪ computed
by extracting the grade-lowered part of the geometric product.

This implementation represents multivectors as objects with 8
↪ components,
indexed by an integer (from 0 to 7) that encodes the basis blade in
↪ binary.
The bit convention is:
    Bit 0 (1) : e1
    Bit 1 (2) : e2
    Bit 2 (4) : e3
Thus:
    0        : scalar
    1 = 0b001 : e1
    2 = 0b010 : e2
    3 = 0b011 : e1e2
    4 = 0b100 : e3
    5 = 0b101 : e1e3
    6 = 0b110 : e2e3
    7 = 0b111 : e1e2e3

All operations assume a Euclidean signature.
"""

import math

# Number of basis vectors (for R^3).
N_BASIS = 3
NV = 1 << N_BASIS  # Total number of basis blades (2^3 = 8)

def popcount(n):
    "Return the number of set bits in the integer n."
    return bin(n).count("1")

def gp_basis(a, b):
    """
    Compute the geometric product of two basis blades represented
    by their integer labels a and b.
    Returns a tuple (c, sign) where c = a XOR b and sign is
    ↪ determined
    by the number of swaps required to interlace the two blades.
    """
    c = a ^ b   # XOR combines the basis indices (removes
    ↪ duplicates).
    sign = 1
    # For each set bit in 'a', count the number of set bits in 'b'
    ↪ in lower positions.
```

```python
            for i in range(N_BASIS):
                if (a >> i) & 1:
                    for j in range(i):
                        if (b >> j) & 1:
                            sign = -sign
        return c, sign

class MultiVector:
    """
    Class representing a multivector in Cl(3,0).
    The components are stored in a list of length 8.
    """
    def __init__(self, coeffs=None):
        if coeffs is None:
            self.coeffs = [0.0] * NV
        else:
            if len(coeffs) != NV:
                raise ValueError("Coefficient list must have length "
                    "{}.".format(NV))
            self.coeffs = [float(c) for c in coeffs]

    def __add__(self, other):
        result = MultiVector()
        for i in range(NV):
            result.coeffs[i] = self.coeffs[i] + other.coeffs[i]
        return result

    def __sub__(self, other):
        result = MultiVector()
        for i in range(NV):
            result.coeffs[i] = self.coeffs[i] - other.coeffs[i]
        return result

    def __mul__(self, other):
        # Scalar multiplication.
        if isinstance(other, (int, float)):
            result = MultiVector()
            for i in range(NV):
                result.coeffs[i] = self.coeffs[i] * other
            return result
        # Geometric product between two multivectors.
        elif isinstance(other, MultiVector):
            result = MultiVector()
            for i in range(NV):
                a_coeff = self.coeffs[i]
                if abs(a_coeff) < 1e-12:
                    continue
                for j in range(NV):
                    b_coeff = other.coeffs[j]
                    if abs(b_coeff) < 1e-12:
                        continue
                    k, sign = gp_basis(i, j)
                    result.coeffs[k] += a_coeff * b_coeff * sign
```

```python
            return result
        else:
            raise TypeError("Unsupported multiplication type with
            ↪ {}".format(type(other)))

    def __rmul__(self, other):
        # Support left multiplication by scalar.
        return self.__mul__(other)

    def __truediv__(self, other):
        # Division by a scalar.
        if isinstance(other, (int, float)):
            result = MultiVector()
            for i in range(NV):
                result.coeffs[i] = self.coeffs[i] / other
            return result
        else:
            raise TypeError("Division only supported by scalars.")

    def __str__(self):
        basis_names = {
            0: "1",
            1: "e1",
            2: "e2",
            3: "e1e2",
            4: "e3",
            5: "e1e3",
            6: "e2e3",
            7: "e1e2e3"
        }
        terms = []
        for i, coeff in enumerate(self.coeffs):
            if abs(coeff) > 1e-10:
                term = "{:.3g}".format(coeff)
                if i != 0:
                    term += basis_names.get(i, "")
                terms.append(term)
        if not terms:
            return "0"
        return " " + ".join(terms)

    def reverse(self):
        """
        Compute the reverse (reversion) of the multivector.
        For a homogeneous grade r element, the factor is
        ↪ (-1)^(r*(r-1)/2).
        This operation is applied componentwise.
        """
        result = MultiVector()
        for i in range(NV):
            if abs(self.coeffs[i]) < 1e-12:
                continue
            r = popcount(i)
```

```python
            factor = (-1)**(r*(r-1)//2)
            result.coeffs[i] = self.coeffs[i] * factor
        return result

    def grade(self):
        """
        Returns the grade of the multivector if it is homogeneous.
        If more than one grade is present, returns None.
        """
        grades = set()
        for i, coeff in enumerate(self.coeffs):
            if abs(coeff) > 1e-10:
                grades.add(popcount(i))
        if len(grades) == 1:
            return grades.pop()
        else:
            return None

    def extract_grade(self, target_grade):
        """
        Extract the part of the multivector with a given grade.
        """
        result = MultiVector()
        for i, coeff in enumerate(self.coeffs):
            if popcount(i) == target_grade:
                result.coeffs[i] = coeff
        return result

    def inverse(self):
        """
        Compute the multiplicative inverse of an invertible blade B.
        For an invertible blade, B~{-1} = B.reverse() / (B *
        ↪  B.reverse())_scalar.
        """
        B_rev = self.reverse()
        prod = self * B_rev
        scalar = prod.coeffs[0]
        if abs(scalar) < 1e-12:
            raise ZeroDivisionError("Inverse does not exist; scalar
                ↪  part is zero.")
        return B_rev / scalar

def basis_vector(i):
    """
    Construct a basis vector.
    For i = 1,2,3, the corresponding basis blade is defined by bit
    ↪  0,1,2 respectively.
    """
    if i < 1 or i > N_BASIS:
        raise ValueError("Basis vector index must be between 1 and
            ↪  {}.".format(N_BASIS))
    mv = MultiVector()
    mv.coeffs[1 << (i - 1)] = 1.0
```

```
        return mv

def inner_product(A, B):
    """
    Compute the inner (contraction) product between two
    ↪   multivectors.
    In this implementation, we assume A is a vector (grade 1) and B
    ↪   is homogeneous of grade r.
    The inner product is defined as the grade (r-1) part of the
    ↪   geometric product A * B.
    """
    grade_B = B.grade()
    if grade_B is None or grade_B < 1:
        raise ValueError("B must be homogeneous of grade >= 1 for a
        ↪   proper inner product.")
    target_grade = grade_B - 1
    prod = A * B
    return prod.extract_grade(target_grade)

def dot(v, B):
    "Alias for inner_product (vector contraction with blade)."
    return inner_product(v, B)

def project_vector(v, B):
    """
    Compute the projection of a vector v onto the subspace defined
    ↪   by an invertible blade B.
    The projection is given by:
        v_parallel = (v · B) * B^{-1}
    where '·' is the inner product (contraction) between v and B.
    """
    return dot(v, B) * B.inverse()

# Demonstration of the projection operator.
if __name__ == "__main__":
    # Construct basis vectors e1, e2, e3.
    e1 = basis_vector(1)
    e2 = basis_vector(2)
    e3 = basis_vector(3)

    # Define a vector v = e1 + 2e2 + 3e3.
    v = e1 + 2 * e2 + 3 * e3
    print("Vector v:")
    print("   ", v)

    # Define a blade B representing the 2-dimensional subspace
    ↪   (plane) spanned by e1 and e2.
    # For a simple blade, we can take the geometric product of two
    ↪   basis vectors:
    B = e1 * e2   # This yields the bivector e1e2.
    print("\nBlade B (e1^e2 plane):")
    print("   ", B)
```

200

```python
# Compute the projection of v onto the subspace defined by B.
v_parallel = project_vector(v, B)
print("\nProjection of v onto the subspace defined by B
 ↪  (v_parallel):")
print("   ", v_parallel)

# For verification, v_parallel should be a vector that lies in
 ↪  the e1-e2 plane.
# (There should be no component along e3.)
```

Chapter 28

Subspace Representation: Blades and Their Geometry

Blades as Algebraic Representatives of Subspaces

Blades are homogeneous multivectors arising from the exterior product of linearly independent vectors. Given a set of independent vectors $v_1, v_2, \ldots, v_r$, the blade defined by

$$B = v_1 \wedge v_2 \wedge \cdots \wedge v_r$$

encapsulates the oriented subspace spanned by these vectors. The grade of the blade, denoted by r, is intrinsically linked to the dimensionality of the subspace. The properties of the wedge product ensure that the blade incorporates information regarding both the magnitude and orientation of the subspace. Its antisymmetric nature excludes redundant contributions, yielding a compact and robust algebraic representative of geometric entities such as lines, planes, and hyperplanes.

Geometric Interpretation of Blades

The geometric significance of blades is revealed by the way they encode both area and orientation. For instance, consider the bivector formed from two non-collinear vectors a and b. The bivector

$$B = a \wedge b$$

represents the oriented area element of the parallelogram defined by a and b. Its magnitude, computed as

$$|B| = \sqrt{B \cdot B},$$

corresponds to the area of the parallelogram, while its direction indicates the plane in which the vectors reside. Higher-grade blades follow an analogous interpretation, with a trivector representing an oriented volume element and so forth. The correspondence between the grade of a blade and the dimensionality of the subspace it represents provides a systematic framework to study complex vectorial relationships via geometric algebra.

Algorithmic Handling of Blades in Computational Systems

The representation and manipulation of blades in computational environments benefit from efficient data structures. Typically, basis blades are encoded as bit masks where each bit represents the inclusion of a particular basis vector. This binary representation allows for rapid determination of the grade by counting the number of set bits. Such an approach is pivotal when establishing the homogeneity of multivectors and ensuring the correctness of subsequent algebraic operations.

An illustrative function for computing the grade of a homogeneous multivector is provided below. This function assumes the multivector is stored as a dictionary, where the keys are integers encoding the basis blades through their binary representation and the values are the corresponding coefficients.

```
def compute_grade(multivector):
    """
    Compute the grade of a given homogeneous multivector.

    The multivector is represented as a dictionary where keys are
       integers,
    each integer encodes a basis blade by its binary representation,
       and the values
    are the corresponding coefficients. The function calculates the
       grade by counting
```

the number of set bits in the key. If the multivector contains
↪ components of
inconsistent grades, a ValueError is raised.

Parameters:
    multivector (dict): A mapping from basis blade codes to
    ↪ coefficients.

Returns:
    int: The grade of the homogeneous multivector.
"""
grades = {bin(blade).count("1") for blade, coeff in
↪ multivector.items() if abs(coeff) > 1e-10}
if len(grades) == 1:
    return grades.pop()
raise ValueError("Multivector components have differing
↪ grades.")
```

The function `compute_grade` utilizes the binary representation of each basis blade to determine the number of vectors contributing to the blade, thereby confirming the subspace dimensionality. Ensuring that multivectors are homogeneous is critical for many computational techniques involving blade operations, as inconsistency in grade may lead to ambiguous or erroneous interpretations in geometric computations.

# Python Code Snippet

```
# Helper function to compute the grade of a homogeneous multivector.
def compute_grade(multivector):
    """
    Compute the grade of a given homogeneous multivector.

    The multivector is represented as a dictionary where keys are
    ↪ integers,
    each integer encodes a basis blade by its binary representation,
    ↪ and the values
    are the corresponding coefficients. The function calculates the
    ↪ grade by counting
    the number of set bits in the key. If the multivector contains
    ↪ components of
    inconsistent grades, a ValueError is raised.

    Parameters:
        multivector (dict): A mapping from basis blade codes to
        ↪ coefficients.

    Returns:

```
        int: The grade of the homogeneous multivector.
    """
    grades = {bin(blade).count("1") for blade, coeff in
    ↪   multivector.items() if abs(coeff) > 1e-10}
    if len(grades) == 1:
        return grades.pop()
    raise ValueError("Multivector components have differing
    ↪   grades.")

# Function to convert a basis blade integer to a human-readable
↪   string.
def blade_to_str(blade, basis_prefix='e'):
    """
    Convert a basis blade represented as an integer bit mask into a
    ↪   string.
    For example, a blade encoded as 5 (binary 101) becomes 'e1^e3'.

    Parameters:
        blade (int): The integer representation of the basis blade.
        basis_prefix (str): The prefix to use for basis vectors.

    Returns:
        str: The string representation of the basis blade.
    """
    indices = []
    idx = 1
    while blade:
        if blade & 1:
            indices.append(f"{basis_prefix}{idx}")
        blade >>= 1
        idx += 1
    return "^".join(indices) if indices else "1"

# Function to compute the wedge product between two basis blades.
def wedge_basis(blade1, blade2):
    """
    Compute the wedge product of two basis blades given by their
    ↪   integer representations.

    The wedge product is defined only if the two blades have no
    ↪   common basis vectors.
    If (blade1 & blade2) is non-zero, the blades share common
    ↪   components and the wedge product
    vanishes. Otherwise, the resulting blade is given by the bitwise
    ↪   XOR of the two blades with
    an appropriate sign determined by the ordering of the basis
    ↪   indices.

    Parameters:
        blade1 (int): The first basis blade (as an integer bit
        ↪   mask).
        blade2 (int): The second basis blade (as an integer bit
        ↪   mask).
```

```
    Returns:
        tuple: (sign, result_blade) where sign is +1 or -1, and
         ↪   result_blade is the integer encoding
             of the resulting blade. If the blades are not
              ↪  disjoint, returns (0, 0).
    """
    # Check for common basis vectors: if any, the wedge product is
     ↪  zero.
    if blade1 & blade2:
        return (0, 0)

    # Function to extract indices from a blade.
    def blade_to_indices(blade):
        indices = []
        pos = 0
        while blade:
            if blade & 1:
                indices.append(pos)
            blade >>= 1
            pos += 1
        return indices

    indices1 = blade_to_indices(blade1)
    indices2 = blade_to_indices(blade2)

    # Compute the number of swaps required to merge indices1 and
     ↪  indices2 into sorted order.
    swaps = 0
    for idx in indices2:
        # Count how many basis vectors from the first blade have an
         ↪   index greater than idx.
        swaps += sum(1 for i in indices1 if i > idx)

    sign = (-1) ** swaps
    result_blade = blade1 ^ blade2
    return (sign, result_blade)

# Function to compute the wedge product of two multivectors.
def wedge_product(mv1, mv2):
    """
    Compute the wedge product of two multivectors.

    Both multivectors are represented as dictionaries that map
     ↪  integer-encoded basis blades
    to their corresponding coefficients. The wedge product is
     ↪  computed by taking all pairwise
    products of components from mv1 and mv2 and combining those
     ↪  using the wedge_basis function.
    Blades that are not disjoint (i.e. share common basis vectors)
     ↪  do not contribute to the result.

    Parameters:
```

```
        mv1 (dict): The first multivector.
        mv2 (dict): The second multivector.

    Returns:
        dict: The resulting multivector after computing the wedge
        ↪    product.
    """
    result = {}
    for blade1, coeff1 in mv1.items():
        for blade2, coeff2 in mv2.items():
            sign, new_blade = wedge_basis(blade1, blade2)
            if sign == 0:
                continue
            new_coeff = coeff1 * coeff2 * sign
            if new_blade in result:
                result[new_blade] += new_coeff
            else:
                result[new_blade] = new_coeff
    return result

# Example usage and testing the functions.
if __name__ == "__main__":
    # Define simple multivectors using bit masks for basis blades.
    # For example, let e1 be 1<<0 (1), e2 be 1<<1 (2), and e3 be
    ↪    1<<2 (4).
    mv_a = {1: 3.0}           # Represents 3*e1
    mv_b = {2: 2.0}           # Represents 2*e2
    mv_c = {4: 1.0}           # Represents 1*e3

    # Compute the wedge product of mv_a and mv_b:
    # Expected result is 3*2 = 6 times the basis blade e1^e2.
    wedge_ab = wedge_product(mv_a, mv_b)
    print("Wedge product of mv_a and mv_b:")
    for blade, coeff in wedge_ab.items():
        print(f"{coeff} * {blade_to_str(blade)}")

    # Compute the wedge product of (mv_a  mv_b) with mv_c:
    # This should yield the volume element e1^e2^e3.
    mv_ab = wedge_ab
    wedge_abc = wedge_product(mv_ab, mv_c)
    print("\nWedge product of (mv_a wedge mv_b) and mv_c:")
    for blade, coeff in wedge_abc.items():
        print(f"{coeff} * {blade_to_str(blade)}")

    # Test the compute_grade function on a homogeneous multivector.
    try:
        grade_ab = compute_grade(mv_ab)
        print(f"\nGrade of mv_a wedge mv_b: {grade_ab}")
    except ValueError as e:
        print(e)
```

Chapter 29

Intersection Operations: Computing Geometric Intersections

Algebraic Framework for Subspace Intersections

Within geometric algebra, subspaces are encoded by blades that inherently capture both dimensionality and orientation. Intersection operations extract the common subspace between distinct geometric entities by leveraging the duality inherent in the algebra. Given a blade A representing one subspace and a blade B representing another, the intersection is obtained by transitioning to the dual space, performing a join via the wedge product, and then returning to the original space. This operation is formally expressed as

$$A \cap B = \left(A^* \wedge B^* \right)^*,$$

where the dual of a blade is defined by

$$A^* = A I^{-1},$$

with I being the pseudoscalar of the algebra. The operation exploits the fact that the union of dual subspaces corresponds to the intersection in the original space.

Intersection via Duality: The Meet Operation

The meet, or intersection, of two subspaces finds its rigorous formulation through the duality transformation. For blades A and B, their duals A^* and B^* are computed using the inverse of the pseudoscalar. The duals embody the complementary subspaces, and their wedge product

$$A^* \wedge B^*$$

forms a blade that represents the join of the complementary subspaces. Reapplying the dual transformation recovers the intersection of the original subspaces. This process is expressed by the formula

$$A \cap B = \Big((A\,I^{-1}) \wedge (B\,I^{-1})\Big) I.$$

This algebraic formulation ensures that the extraction of common components is both mathematically rigorous and computationally efficient.

An illustrative implementation of this process is provided by the function below. The function accepts the blades corresponding to the subspaces along with the pseudoscalar and its inverse. It then computes the duals, aggregates them using the wedge product, and applies the reverse dual transformation. The resulting blade precisely characterizes the intersection of the two subspaces.

```
def compute_meet(bladeA, bladeB, pseudoscalar, inv_pseudoscalar):
    """
    Compute the intersection (meet) of two subspaces represented by
      blades using
    the duality principle in geometric algebra.

    The intersection is calculated using the formula:

        meet = dual( wedge( dual(bladeA), dual(bladeB) ) )
             = ( (bladeA * inv_pseudoscalar) ^ (bladeB *
               inv_pseudoscalar) ) * pseudoscalar

    where '^' denotes the wedge product and '*' denotes the
      geometric product.
    The pseudoscalar and its inverse must be provided.

    Parameters:
        bladeA (multivector): The blade representing the first
          subspace.
```

```
    bladeB (multivector): The blade representing the second
    ↪ subspace.
    pseudoscalar (multivector): The pseudoscalar of the algebra.
    inv_pseudoscalar (multivector): The inverse of the
    ↪ pseudoscalar.

Returns:
    multivector: The blade representing the intersection of the
    ↪ two subspaces.
"""
dualA = geometric_product(bladeA, inv_pseudoscalar)
dualB = geometric_product(bladeB, inv_pseudoscalar)
joined_duals = wedge_product(dualA, dualB)
return geometric_product(joined_duals, pseudoscalar)
```

Computational Considerations in Intersection Operations

The implementation of intersection operations requires precise management of multivector components and adherence to the algebraic structure of geometric algebra. The meet operation described above relies on the homogeneity of each blade, as any variation in grade across the components would obscure the interpretation of an intersection. The computational procedure is inherently modular: dual transformations, wedge products, and the final reapplication of the pseudoscalar collectively produce a robust algorithm for determining common subspaces.

In practice, numerical stability when computing the inverse of the pseudoscalar is critical; errors here can propagate through the dual transformations and adversely affect the outcome. The antisymmetry of the wedge product ensures that any overlapping contributions from non-disjoint elements are automatically nullified, making the approach well suited for high-dimensional applications. The explicit separation of each computational stage facilitates both debugging and optimization of the algorithm in computational implementations.

Python Code Snippet

```
# Import necessary libraries from the clifford module for geometric
↪ algebra operations
from clifford import Cl
import numpy as np
```

```python
# Create a 3-dimensional geometric algebra (Cl(3,0)) instance.
# This sets up basis vectors e1, e2, e3, and the pseudoscalar.
layout, blades = Cl(3)
locals().update(blades)

# Retrieve the pseudoscalar and compute its inverse.
I = layout.pseudoScalar
inv_I = I.inv()

def geometric_product(A, B):
    """
    Compute the geometric product of two multivectors A and B.
    In the 'clifford' library the '*' operator already implements
    ↪ the
    geometric product.

    Parameters:
        A (multivector): The first multivector.
        B (multivector): The second multivector.

    Returns:
        multivector: The geometric product A*B.
    """
    return A * B

def wedge_product(A, B):
    """
    Compute the outer (wedge) product of two multivectors A and B.
    The '^' operator provided by 'clifford' performs the exterior
    ↪ product.

    Parameters:
        A (multivector): The first multivector.
        B (multivector): The second multivector.

    Returns:
        multivector: The wedge product A^B.
    """
    return A ^ B

def compute_meet(bladeA, bladeB, pseudoscalar, inv_pseudoscalar):
    """
    Compute the intersection (meet) of two subspaces represented by
    ↪ blades
    using the duality principle in geometric algebra.

    The intersection is calculated using the formula:

        meet = dual( wedge( dual(bladeA), dual(bladeB) ) )
             = ((bladeA * inv_pseudoscalar) ^ (bladeB *
               ↪ inv_pseudoscalar)) * pseudoscalar
```

```
where:
    - dual(blade) is computed as blade * inv_pseudoscalar,
    - the '^' operator stands for the wedge (exterior) product,
    - and the final multiplication with the pseudoscalar yields
      the dual of
        the joined duals, recovering the intersection of the
        original subspaces.

Parameters:
    bladeA (multivector): Blade representing the first subspace.
    bladeB (multivector): Blade representing the second
      subspace.
    pseudoscalar (multivector): The pseudoscalar of the
      geometric algebra.
    inv_pseudoscalar (multivector): The inverse of the
      pseudoscalar.

Returns:
    multivector: The blade representing the intersection (meet)
      of the two subspaces.
"""
# Compute the duals of bladeA and bladeB.
dualA = geometric_product(bladeA, inv_pseudoscalar)
dualB = geometric_product(bladeB, inv_pseudoscalar)

# Calculate the join of the dual subspaces via the wedge
  product.
joined_duals = wedge_product(dualA, dualB)

# Recover the intersection of the original subspaces by taking
  the dual of the join.
meet = geometric_product(joined_duals, pseudoscalar)
return meet

if __name__ == "__main__":
    # Example: Define two subspaces in 3D space.
    # Here, we represent two planes whose intersection forms a line.

    # Define plane1 as the blade spanning the e1 and e2 directions.
    plane1 = e1 ^ e2   # Represents a plane in the e1-e2 plane.

    # Define plane2 as the blade spanning the e2 and e3 directions.
    plane2 = e2 ^ e3   # Represents a plane in the e2-e3 plane.

    # Compute their intersection (meet) using the duality-based
      algorithm.
    intersection = compute_meet(plane1, plane2, I, inv_I)

    # Output the resulting blades.
    print("Plane 1 (blade):", plane1)
    print("Plane 2 (blade):", plane2)
    print("Intersection (meet):", intersection)
```

Chapter 30

Dirac Algebra in GA: Bridging Quantum Theory

Algebraic Foundations

The Dirac algebra finds a natural reformulation within the framework of geometric algebra. In the standard representation, the Dirac algebra is associated with the Clifford algebra $Cl(1,3)$, which corresponds to the four-dimensional Minkowski spacetime. The basis elements of this algebra obey the relation

$$\gamma_\mu \gamma_\nu + \gamma_\nu \gamma_\mu = 2\eta_{\mu\nu},$$

where γ_μ are the gamma matrices and $\eta_{\mu\nu}$ is the Minkowski metric of signature $(+,-,-,-)$. Within geometric algebra, these gamma matrices are reinterpreted as vectors of the algebra, and the geometric product naturally encodes both symmetric and antisymmetric parts of their multiplication. This correspondence provides an elegant algebraic structure that unifies the operations of metric contraction and exterior product under a single multiplication rule.

Representation of Gamma Matrices and Spinors

The identification of the gamma matrices with basis vectors in $Cl(1,3)$ allows for a reinterpretation of Dirac spinors. In the geometric algebra formulation, spinors are represented as multivectors, often elements of the even subalgebra, which is isomorphic to the Pauli algebra in three dimensions. This viewpoint renders the underlying geometric meaning of spinor transformations transparent. The traditional matrix representation is replaced by a formulation in which rotations, boosts, and reflections emerge from the geometric product. In particular, the operations that previously required cumbersome matrix algebra now appear as natural geometric transformations. The algebraic structure thereby exhibits the intrinsic connection between the symmetry properties of spacetime and the behavior of quantum spin.

Implications for Quantum Physics

Recasting the Dirac algebra in terms of geometric algebra has notable implications for the interpretation of quantum theory. The Dirac equation, when expressed in this formalism, possesses a geometric clarity that demystifies many of its longstanding conceptual challenges. The Dirac operator in geometric algebra is constructed as a combination of the derivative operator and the multivector representation of the spinor field. When the derivative operator, denoted by ∇, is combined via the geometric product with the spinor, the resulting formulation reveals symmetry properties that are obscured in the conventional approach. The interplay between the geometric product and the exterior product (wedge product) codifies the transformation properties of quantum operators and simplifies manipulations that involve conservation laws and invariants. Consequently, the geometric algebra formulation enhances computational efficiency and offers deeper insight into the geometry underlying the Dirac theory.

Computational Implementation: Python Function for the Dirac Operator in GA

A computational representation of the Dirac operator within the geometric algebra framework can be encapsulated in a single Python function. The function computes the action of the Dirac operator by performing the geometric product between the derivative operator and a multivector representing the Dirac spinor. The derivative operator, structured as a multivector, encodes the partial derivatives with respect to spacetime coordinates. The following function demonstrates this computation:

```
def ga_dirac_operator(spinor, deriv):
    """
    Compute the action of the Dirac operator in a geometric algebra
    ↪ framework.

    The computation involves obtaining the geometric product of the
    ↪ derivative operator
    with the Dirac spinor. In this context, 'spinor' is the
    ↪ multivector representation
    of the Dirac spinor, and 'deriv' encodes the derivative operator
    ↪ as a multivector,
    which includes the partial derivatives with respect to spacetime
    ↪ coordinates.

    Parameters:
        spinor (multivector): The multivector representing the Dirac
        ↪ spinor.
        deriv (multivector): The derivative operator represented as
        ↪ a multivector.

    Returns:
        multivector: The result of applying the Dirac operator via
        ↪ the geometric product.
    """
    return deriv * spinor
```

Python Code Snippet

```
# Import required libraries
import numpy as np
from clifford import Cl

# Create a Clifford Algebra for Minkowski spacetime (signature: +,
↪ -, -, -)
```

```
layout, blades = Cl(1, 3)

# Define the basis vectors (gamma matrices) corresponding to
↪ spacetime directions.
# These satisfy the Clifford relation:
#   gamma_mu * gamma_nu + gamma_nu * gamma_mu = 2 * eta_{mu nu},
# where eta is the Minkowski metric: diag(1, -1, -1, -1).
gamma0 = blades['e0']
gamma1 = blades['e1']
gamma2 = blades['e2']
gamma3 = blades['e3']


# Define a sample Dirac spinor as a multivector.
# In the GA formulation, Dirac spinors are typically represented as
↪ elements
# of the even subalgebra. Here, we choose a simple example combining
↪ a scalar
# and a bivector part.
psi = 1.0 + gamma1 * gamma2

# -----------------------------------------------------------------
# Numerical derivative function for approximating partial
↪ derivatives.
def numerical_derivative(func, x, h=1e-5):
    """
    Approximates the derivative of a function 'func' at the point x
    ↪ using
    central difference.
    """
    return (func(x + h) - func(x - h)) / (2 * h)

# -----------------------------------------------------------------
# Define a scalar field component of the Dirac spinor.
# This function mimics the spatial-temporal variation of the
↪ spinor's
# scalar part (e.g., a Gaussian distribution).
def psi_scalar_field(t, x, y, z):
    return np.exp(- (t**2 + x**2 + y**2 + z**2))

# -----------------------------------------------------------------
# Construct the derivative operator (nabla) in GA.
# The derivative operator in Minkowski space is given by:
#   ∇ = gamma0 * ∂/∂t + gamma1 * ∂/∂x + gamma2 * ∂/∂y + gamma3 * ∂/∂z
def ga_derivative_operator(coords, field_func):
    """
    Computes the GA derivative operator at a given spacetime
    ↪ coordinate.

    Parameters:
        coords      : tuple (t, x, y, z) representing the spacetime
        ↪ point.
        field_func: a function representing a scalar field from
        ↪ which to
```

```
        compute numerical partial derivatives.

    Returns:
        A multivector representing the derivative operator at the
        ↪ given point.
    """
    t, x, y, z = coords

    # Compute numerical partial derivatives for each coordinate
    ↪ direction.
    dt = numerical_derivative(lambda t_val: field_func(t_val, x, y,
    ↪ z), t)
    dx = numerical_derivative(lambda x_val: field_func(t, x_val, y,
    ↪ z), x)
    dy = numerical_derivative(lambda y_val: field_func(t, x, y_val,
    ↪ z), y)
    dz = numerical_derivative(lambda z_val: field_func(t, x, y,
    ↪ z_val), z)

    # Assemble the derivative operator as a multivector.
    deriv = dt * gamma0 + dx * gamma1 + dy * gamma2 + dz * gamma3
    return deriv

# -----------------------------------------------------------------
# Define the GA Dirac operator.
# In this formulation, the Dirac operator is given by the geometric
↪ product
# of the derivative operator with the Dirac spinor:
#   D =
def ga_dirac_operator(spinor, deriv):
    """
    Applies the Dirac operator (constructed via the geometric
    ↪ product) to
    a Dirac spinor.

    Parameters:
        spinor: multivector representation of the Dirac spinor.
        deriv : multivector representation of the derivative
        ↪ operator.

    Returns:
        multivector: the result of the Dirac operator acting on the
        ↪ spinor.
    """
    return deriv * spinor

# -----------------------------------------------------------------
# Demonstration of the GA Dirac operator

# Choose a sample spacetime coordinate (t, x, y, z)
coords = (0.1, 0.2, 0.3, 0.4)
```

```
# Compute the derivative operator at the chosen coordinate using the
↪  scalar field.
deriv_op = ga_derivative_operator(coords, psi_scalar_field)

# Compute the action of the Dirac operator on the Dirac spinor psi.
result = ga_dirac_operator(psi, deriv_op)

# Display the resulting multivector, which encapsulates both the
↪  derivative
# action and the intrinsic geometric products in the Dirac algebra.
print("Result of the Dirac operator acting on the Dirac spinor:")
print(result)
```

Chapter 31

Reformulating Maxwell's Equations with GA: A New Perspective

Geometric Algebra Formulation of Electromagnetism

Maxwell's equations, traditionally expressed as a system of four differential equations, can be recast into a remarkably compact form using geometric algebra. In this formulation the electromagnetic field is represented by the bivector

$$F = \mathbf{E} + I\mathbf{B},$$

where $\mathbf{E}$ and $\mathbf{B}$ denote the electric and magnetic field vectors, respectively, and I is the unit pseudoscalar of Minkowski spacetime. This bivector encapsulates the antisymmetric nature of the conventional electromagnetic field tensor. The geometric product unifies the inner (dot) and outer (wedge) products, thereby incorporating both divergence and curl operations into a single algebraic operation. This intrinsic unification simplifies both the conceptual understanding and subsequent computational treatment of electromagnetic phenomena.

Unified Expression of Maxwell's Equations in GA

Within the geometric algebra framework the four classical Maxwell equations merge into one compact statement. Denoting the spacetime derivative operator by

$$\nabla,$$

the complete set of electromagnetic laws is described by the equation

$$\nabla F = J,$$

where J is the four-current vector. The geometric product ∇F naturally separates into components that correspond to the conventional divergence and curl. In the absence of sources, the condition $\nabla F = 0$ combines the source-free Maxwell equations into a single relation. The unified treatment demonstrates that the apparent complexity of classical electromagnetism reflects only the limitations of conventional vector calculus and not a fundamental physical intricacy. In geometric algebra, the coordinated split between inner and outer products automatically yields the scalar and bivector parts that represent, respectively, the inhomogeneous and homogeneous Maxwell laws.

Computational Implementation in Python

The clarity of the GA reformulation of Maxwell's equations translates directly into computational efficiency. A typical implementation in Python involves constructing the electromagnetic bivector from given electric and magnetic field vectors and then applying the GA derivative operator. The following function exemplifies the process. It takes as input the multivector representation of the derivative operator, the electric field $\mathbf{E}$, and the magnetic field $\mathbf{B}$, along with the unit pseudoscalar I. The electromagnetic bivector is assembled as

$$F = \mathbf{E} + I\mathbf{B},$$

and its derivative is computed via the geometric product. This function encapsulates the unified form of Maxwell's equations in a succinct computational module.

```python
def compute_ga_maxwell(deriv, E, B, I):
    """
    Compute the geometric algebra representation of Maxwell's
    ↪ equations.

    Constructs the electromagnetic bivector F from the electric
    ↪ field E
    and the magnetic field B, with F defined as F = E + I * B. The
    derivative operator 'deriv' is then applied to F via the
    ↪ geometric
    product to yield the GA form of Maxwell's equations.

    Parameters:
        deriv: Multivector representing the spacetime derivative
        ↪ operator.
        E    : Vector representing the electric field.
        B    : Vector representing the magnetic field.
        I    : Unit pseudoscalar of the spacetime algebra.

    Returns:
        Multivector representing the derivative of the
        ↪ electromagnetic field,
        capturing the unified GA formulation of Maxwell's equations.
    """
    # Construct the electromagnetic bivector.
    F = E + I * B
    return deriv * F
```

Python Code Snippet

```python
import numpy as np
from clifford import Cl

# Create a Minkowski spacetime algebra with signature (1, -1, -1,
↪ -1).
# Cl(1, 3) sets up the Clifford algebra for Minkowski spacetime.
layout, blades = Cl(1, 3, names='gamma')
locals().update(blades)

# Define the unit pseudoscalar (I) for the algebra.
I = layout.pseudo_scalar

def compute_ga_maxwell(deriv, E, B, I):
    """
    Compute the geometric algebra representation of Maxwell's
    ↪ equations.

    Constructs the electromagnetic bivector F from the electric
    ↪ field E
```

```
    and the magnetic field B, with F defined as F = E + I * B. The
    derivative operator 'deriv' is then applied to F via the
    ↪ geometric
    product to yield the GA form of Maxwell's equations.

    Parameters:
        deriv: Multivector representing the spacetime derivative
        ↪ operator.
        E    : Vector representing the electric field.
        B    : Vector representing the magnetic field.
        I    : Unit pseudoscalar of the spacetime algebra.

    Returns:
        Multivector representing the derivative of the
        ↪ electromagnetic field,
        capturing the unified GA formulation of Maxwell's equations.
    """
    # Construct the electromagnetic bivector F
    F = E + I * B
    # Apply the derivative operator using the geometric product.
    return deriv * F


# --------------------------------------------------
# Demonstration of the GA Formulation of Maxwell's Equations
# --------------------------------------------------
# In geometric algebra, the electromagnetic field is represented as:
#       F = E + I * B
# where E is the electric field vector, B is the magnetic field
↪ vector,
# and I is the unit pseudoscalar of the Minkowski spacetime.
#
# The unified Maxwell's equations can be written as:
#       nabla * F = J
# where nabla (deriv) is the spacetime derivative operator and J is
↪ the four-current.
#
# For this example, we simplify the derivative operator.
# In a full treatment, nabla = gamma^mu (/x^mu).
# Here, we use gamma0 as a placeholder representing the time
↪ derivative component.
deriv = gamma0

# Define arbitrary constant vector fields for the electric and
↪ magnetic fields.
# These values can be replaced with space-time varying fields in a
↪ full simulation.
E = 1.0 * gamma1 + 2.0 * gamma2 + 3.0 * gamma3    # Electric field
↪ vector
B = 0.5 * gamma1 - 1.0 * gamma2 + 0.0 * gamma3    # Magnetic field
↪ vector

# Assemble the electromagnetic bivector F.
F = E + I * B
```

```
# Compute the derivative of F using the GA formulation of Maxwell's
↪   equations.
result = compute_ga_maxwell(deriv, E, B, I)

# Print the electromagnetic bivector and the result of applying the
↪   derivative operator.
print("Electromagnetic bivector F =", F)
print("Result of derivative operator on F =", result)
```

Chapter 32

Equations of Motion: GA in Dynamical Systems

Formulating Dynamics in Geometric Algebra

Geometric algebra provides a natural framework for reformulating classical dynamics through the language of multivectors and geometric products. In this approach, positions, velocities, and forces are all expressed as elements of a unified algebraic structure. By representing kinematic and dynamic quantities as multivectors, the scalar, vector, and higher-grade components interact via the geometric product. The resulting equations of motion capture both magnitude and orientation in a single compact expression. For example, the conventional second-order differential equations describing a particle's trajectory can be reformulated into first-order multivector equations, where the time derivative of the state yields an amalgam of inertial effects and external influences.

The geometric derivative operator, often denoted by ∇, unifies both divergence and curl operations. When applied to a multivector-valued function of time, it produces a derivative that naturally decomposes into parts corresponding to conservative forces and inertial rotations. This intrinsic decomposition simplifies the separation of dissipative effects from energy-conserving dynamics. The integrated structure inherent in geometric algebra allows for a more unified treatment of translational and rotational motion, which is of particular relevance in the study of rigid body dynamics and fields.

The Geometric Derivative and Differential Equations

The GA formulation of dynamics replaces traditional vector calculus by employing the geometric product to define derivatives and integrals. In this setting, a dynamical state is represented by a multivector $X(t)$ whose time derivative is computed as $\dot{X}(t)$. The equations of motion take the form

$$\dot{X}(t) = \nabla X(t),$$

where the operator ∇ acts in a manner that encompasses both divergence-like and rotation-like effects. This formulation has the advantage of automatically encoding conservation laws. The decomposition of $\nabla X(t)$ into scalar, vector, and bivector components parallels the physical decomposition into energy, momentum, and angular momentum, respectively. Moreover, the algebraic structure eschews the need for coordinate-specific manipulations, providing an invariant description of the dynamics.

A key aspect of this approach is the reduction of the conventional second-order differential equations into first-order equations using the properties of the geometric product. This shift in perspective streamlines the process of deriving motion equations in complex systems, particularly when multiple degrees of freedom are involved. The combined utility of the geometric product in handling both inner and outer products yields a unified formalism that is advantageous for both analytical derivation and numerical simulation.

Application to Rigid Body Dynamics

In the domain of rigid body dynamics, the orientation and rotation of an object are conveniently encoded by rotors, which are even-grade multivectors that transform vectors via sandwiching. The time evolution of the rotor, representing the body's orientation, is governed by a differential equation that encapsulates not only the angular velocity but also the applied torques. In this GA formulation, angular momentum and torque couple naturally through the geometric product, thereby simplifying the treatment of conservation laws and symmetry operations.

The dynamics of a rigid body can be described using an equation of motion that includes both translational and rotational degrees of freedom. For the rotational part, the evolution of a rotor $R(t)$ is determined by

$$\dot{R}(t) = \frac{1}{2} \Omega R(t),$$

where Ω is a bivector representing the angular velocity. The factor $\frac{1}{2}$ arises from the bivector nature of Ω under the rotor representation. This formulation allows for an elegant treatment of rotations in any number of dimensions and facilitates the integration of the equations of motion without resorting to coordinate-dependent parametrizations such as Euler angles or quaternions.

In addition, the interplay between translational motion and rotation becomes simpler to analyze when the full state of the system is encapsulated in a single multivector. The geometric algebra framework permits the direct incorporation of forces and torques into a unified equation, thereby bridging the gap between Newtonian mechanics and modern geometric techniques. The homogeneous treatment of scalar and bivector components provides a clear path for deriving invariants of motion that are directly related to physical conservation laws.

Computational Implementation for GA-based Dynamics

The practical realization of dynamical simulations using geometric algebra relies on efficient computational methods. A common strategy is to employ numerical integration schemes that update the multivector state over discrete time increments. The simplicity of the GA form of the equations of motion permits the use of straightforward integration methods, such as the forward Euler scheme, while still retaining the intrinsic geometric properties of the state variables.

The Python function presented below embodies a minimalistic implementation of a forward Euler integrator tailored for a dynamical system defined in a geometric algebra framework. This function accepts the current state, a timestep, and a derivative function that encodes the GA-based differential equations. The resulting state is updated according to the equation

$$X(t + \Delta t) = X(t) + \Delta t \cdot \dot{X}(t),$$

where $\dot{X}(t)$ is computed via the provided derivative function.

```
def integrate_eom(state, dt, derivative_func):
    """
    Compute the subsequent state of a dynamical system governed by
    ↪ GA-based equations of motion.

    This function accepts the current state represented as a
    ↪ multivector and advances it
    by a timestep dt using a provided derivative function that
    ↪ encapsulates the GA dynamics.
    The state update is computed using a simple forward Euler
    ↪ integration scheme,
    which is suitable for illustrative purposes in a GA framework.

    Parameters:
        state : Multivector representing the current dynamical state
        ↪ (e.g., position, momentum).
        dt    : Floating point scalar indicating the integration
        ↪ timestep.
        derivative_func : Function that computes the GA derivative
        ↪ of the system state.

    Returns:
        Multivector representing the updated state after time dt.
    """
    return state + dt * derivative_func(state)
```

Python Code Snippet

```
"""
Python code snippet for GA-based equations of motion and rotor
↪ dynamics using
forward Euler integration. This script demonstrates how to simulate:
  1. A general linear dynamical system expressed as dX/dt = A * X,
  ↪ where X is a multivector
     and A is a constant multivector operator.
  2. Rigid body (rotor) dynamics expressed as dR/dt = 0.5 * omega *
  ↪ R, where R is a rotor (even-grade multivector)
     and omega is a constant bivector representing the angular
     ↪ velocity.
```

```python
The script uses the 'clifford' package to implement geometric
    algebra operations.
Make sure to install it via: pip install clifford
"""

import numpy as np
from clifford import Cl

# Initialize a 3D geometric algebra (Clifford algebra) space.
layout, blades = Cl(3)

# =======================================================
# Forward Euler Integrator for GA-Based Dynamics
# =======================================================
def integrate_eom(state, dt, derivative_func):
    """
    Compute the subsequent state of a dynamical system governed by
    GA-based equations of motion using the forward Euler method.

    Parameters:
        state           : Multivector representing the current state
            (e.g., position, orientation).
        dt              : Floating point scalar indicating the
            timestep.
        derivative_func: Function that computes the derivative of
            the state.

    Returns:
        Multivector representing the updated state after time
            increment dt.
    """
    return state + dt * derivative_func(state)

# =======================================================
# Example 1: Linear Dynamical System
# =======================================================
# Define a constant multivector A for the linear system.
# The corresponding equation of motion is: dX/dt = A * X.
A = 0.2 + 0.1 * blades['e1']    # Example: a combination of scalar and
    vector parts.

def linear_derivative(X):
    """
    Compute the derivative of the multivector state X for a linear
        system.

    Returns:
        The time derivative dX/dt computed as the geometric product
            A * X.
    """
    return A * X

# =======================================================
```

```python
# Example 2: Rotor Dynamics in Rigid Body Motion
# =====================================================
# Define a constant bivector omega that represents the angular
↪ velocity.
# The rotor (orientation) evolution is governed by: dR/dt = 0.5 *
↪ omega * R.
omega = 0.5 * blades['e12'] + 0.3 * blades['e23']  # Example
↪ bivector for angular velocity.

def rotor_derivative(R):
    """
    Compute the derivative of the rotor R representing orientation.

    Returns:
        The time derivative dR/dt computed as 0.5 * (omega * R).
    """
    return 0.5 * (omega * R)


# =====================================================
# Simulation Parameters and Initialization
# =====================================================
T = 1.0         # Total simulation time (seconds)
dt = 0.01       # Time step (seconds)
num_steps = int(T / dt)

# Initial condition for the linear system: X could represent a
↪ position multivector.
X = 1.0 + blades['e2']  # Example initial state combining scalar and
↪ vector part.

# Initial condition for rotor dynamics: R is a unit rotor
↪ representing no initial rotation.
R = 1.0  # A pure scalar rotor, satisfying R * ~R = 1 by default.

# Lists to store the history of the states for later analysis or
↪ visualization.
X_history = []
R_history = []

# =====================================================
# Time Integration Loop using Forward Euler Method
# =====================================================
for step in range(num_steps):
    # Record the current states.
    X_history.append(X)
    R_history.append(R)

    # Update the linear dynamical state (e.g., position, momentum).
    X = integrate_eom(X, dt, linear_derivative)

    # Update the rotor state (orientation) for rigid body dynamics.
    R = integrate_eom(R, dt, rotor_derivative)
```

```
# ===================================================
# Output the Final States
# ===================================================
print("Final state X (linear dynamics):")
print(X)

print("\nFinal rotor R (orientation state):")
print(R)
```

Chapter 33

GA in Quantum Mechanics: A Unified Language

The Algebraic Structure of Quantum States

In conventional quantum mechanics, states are typically represented as vectors in a complex Hilbert space. Geometric algebra (GA) reformulates this notion by expressing quantum states as multivectors, thus unifying traditional complex amplitudes and spatial orientation into a single algebraic entity. In this framework, spinors become even-grade multivectors that naturally combine aspects of magnitude and phase, and provide an intrinsic geometric interpretation. A general quantum state may be written as

$$\Psi = \psi_0 + \psi_1 e_1 + \psi_{12} e_{12} + \psi_{123} e_{123} + \cdots,$$

where each component ψ_k carries information analogous to the standard probability amplitudes, yet the entire expression is embedded in a coordinate-free algebra. This representation affords an invariant description under rotations and reflections, meaning that the state transformation laws can be expressed directly via the geometric product. The formalism thus eliminates the need for an extraneous complex space by merging the algebra of rotations with the algebra of scalar quantities.

Observables and Operators in Geometric Algebra

In the GA approach, physical observables are encoded as multivector operators acting on the state multivectors. Elements that conventionally correspond to matrices, such as the Pauli matrices or Dirac gamma matrices, are reinterpreted as specific bivectors or higher-grade constructs within the algebra. For instance, the operator corresponding to spin can be written in terms of bivectors that generate rotations in the respective subspaces. The action of an observable O on a state Ψ is expressed by the geometric product

$$O\Psi,$$

which, due to the distributive and associative properties of the geometric product, naturally combines the effects of projection and rotation. In addition, inner products are redefined in GA to take into account the full multivector structure, allowing a seamless computation of expectation values. Such redefinitions offer a unified language that avoids layered abstractions and provides a more intuitive geometric interpretation of measurement processes.

Quantum Dynamics and the Geometric Product

The time evolution of a quantum state in conventional formulations is governed by the Schrödinger equation. Within the GA framework, the dynamics are reformulated in terms of the geometric product and the exponential of multivectors. Consider a Hamiltonian operator H that is expressed as a multivector; the time evolution operator may then be written as

$$U(t) = \exp\left(-\frac{i}{\hbar}Ht\right),$$

where the imaginary unit i is now interpreted as a bivector that effects a rotation in the appropriate subspace. This exponentiation is computed using the power series expansion adapted to the geometric product, which inherently treats scalar and bivector components on an equal footing. The resulting equation

$$\frac{d\Psi(t)}{dt} = -\frac{1}{\hbar}\left(IH\right)\Psi(t)$$

encapsulates both the phase evolution and the dynamical rotation of the state. Here, the bivector I serves a role analogous to the conventional imaginary unit but carries geometrical significance, facilitating the analysis of phenomena such as spin precession and entanglement using an invariant algebraic structure.

Computational Considerations in GA Quantum Mechanics

The computational implementation of quantum mechanics through geometric algebra yields significant advantages. By consolidating scalar, vector, and higher-grade operations into a single unified product, the necessity for multiple coordinate systems and transformation rules is removed. This consolidation translates into a reduction in computational overhead and complexity when simulating quantum systems. Numerical algorithms can operate directly on multivector states using well-defined operations such as the geometric product and its corresponding derivatives.

For example, a Python function that computes the time derivative of a quantum state governed by a GA-formulated Hamiltonian would encapsulate the core dynamical rules in a concise, algorithmic form. The function accepts a state multivector and outputs its derivative through the geometric product with the Hamiltonian operator. Although the specific programming code will be provided in a subsequent section, the structure of such a function typically adheres to a single-responsibility principle in order to maintain clarity and efficiency in the simulation. This approach promotes code that mirrors the elegance of the underlying mathematics.

The unified treatment of quantum states and observables in GA not only simplifies theoretical derivations but also enables the development of computational frameworks that are both efficient and robust. The geometric interpretation helps avoid coordinate singularities and preserves the symmetry properties of the physical system, which is particularly advantageous in simulations of entangled or highly correlated quantum states.

Python Code Snippet

```
import math
```

```python
def permutation_sign(seq):
    """
    Compute the sign of the permutation required to sort the
    ↪ sequence.
    For each pair (i, j) with i < j, if seq[i] > seq[j] then a swap
    ↪ is needed.
    Equal elements do not contribute to sign change.
    """
    sign = 1
    n = len(seq)
    for i in range(n):
        for j in range(i+1, n):
            if seq[i] > seq[j]:
                sign *= -1
    return sign

def blade_product(blade1, blade2):
    """
    Compute the geometric product of two basis blades.

    Each blade is represented as a tuple of integers (e.g., (1,),
    ↪ (1,2), etc.).
    The product is found by concatenating the two blades, computing
    ↪ the sign
    required to reorder them into non-decreasing order, and then
    ↪ canceling duplicate
    indices (since e_i^2 = 1 in an Euclidean space).

    Returns (sign, blade) where blade is the canonical sorted tuple
    ↪ after cancellations.
    """
    combined = list(blade1) + list(blade2)
    # Compute sign by counting inversions needed to order the
    ↪ combined list
    s = permutation_sign(combined)
    # Count occurrences of each index; basis elements appear mod 2
    ↪ (even duplicates cancel)
    counts = {}
    for idx in combined:
        counts[idx] = counts.get(idx, 0) + 1
    # Build the canonical blade: only include indices with an odd
    ↪ count
    result = []
    for idx in sorted(counts.keys()):
        if counts[idx] % 2 == 1:
            result.append(idx)
    return s, tuple(result)

class Multivector:
    """
    A simple implementation of multivectors in a geometric algebra.
```

*The multivector is stored as a dictionary mapping basis blades
↪ (represented as
tuples of integers) to coefficients. The scalar is represented
↪ by the empty tuple ().*
"""
```python
def __init__(self, coeffs=None):
    # coeffs: dictionary where key is a tuple representing the
    ↪ basis blade, value is coefficient
    if coeffs is None:
        self.coeffs = {}
    else:
        # Remove near-zero coefficients for cleanliness.
        self.coeffs = {blade: val for blade, val in
            ↪ coeffs.items() if abs(val) > 1e-12}

def __add__(self, other):
    result = self.coeffs.copy()
    for blade, val in other.coeffs.items():
        result[blade] = result.get(blade, 0) + val
    return Multivector(result)

def __sub__(self, other):
    result = self.coeffs.copy()
    for blade, val in other.coeffs.items():
        result[blade] = result.get(blade, 0) - val
    return Multivector(result)

def __rmul__(self, scalar):
    # Right multiplication by a scalar.
    if isinstance(scalar, (int, float)):
        return Multivector({blade: scalar * val for blade, val
            ↪ in self.coeffs.items()})
    else:
        raise NotImplementedError("Right multiplication not
            ↪ defined for type " + str(type(scalar)))

def __mul__(self, other):
    if isinstance(other, Multivector):
        # Geometric product: multiply each term and sum the
        ↪ contributions.
        result = {}
        for blade1, val1 in self.coeffs.items():
            for blade2, val2 in other.coeffs.items():
                sign, blade = blade_product(blade1, blade2)
                result[blade] = result.get(blade, 0) + val1 *
                    ↪ val2 * sign
        return Multivector(result)
    elif isinstance(other, (int, float)):
        # Scalar multiplication.
        return Multivector({blade: val * other for blade, val in
            ↪ self.coeffs.items()})
    else:
```

```python
            raise NotImplementedError("Multiplication with type " +
            ↪   str(type(other)) + " not implemented.")

    def __str__(self):
        if not self.coeffs:
            return "0"
        terms = []
        # Sort by grade (length of blade) and then
        ↪   lexicographically.
        for blade, val in sorted(self.coeffs.items(), key=lambda x:
        ↪   (len(x[0]), x[0])):
            if blade == ():
                term = f"{val:.3g}"
            else:
                # Represent basis blade e_i1e_i2...
                blade_str = "".join([f"e{idx}" for idx in blade])
                term = f"{val:.3g}{blade_str}"
            terms.append(term)
        return " + ".join(terms)

    def exp(self, order=20):
        """
        Compute the exponential of the multivector using a power
        ↪   series expansion:

        exp(A) = 1 + A + (1/2!) A^2 + (1/3!) A^3 + ... up to the
        ↪   given number of terms.
        """
        result = Multivector({(): 1.0})
        term = Multivector(self.coeffs.copy())
        for n in range(1, order):
            coeff = 1.0 / math.factorial(n)
            result = result + coeff * term
            term = term * self
        return result

def time_derivative(Psi, H, I, hbar=1.0):
    """
    Compute the time derivative of the quantum state in geometric
    ↪   algebra,
    according to the equation:

    d/dt = -(1/)*(I * H) *

    where I is the pseudoscalar acting analogously to the imaginary
    ↪   unit.
    """
    return -(1/hbar) * (I * H) * Psi

def time_evolution_operator(H, I, t, hbar=1.0, order=20):
    """
    Compute the time evolution operator U(t) using the GA
    ↪   formulation:
```

```
        U(t) = exp(- (I * H * t) / )

    The exponential is computed via a power series expansion.
    """
    U = ((-t/hbar) * (I * H)).exp(order=order)
    return U

# Helper functions to easily create scalars and basis vectors.
def scalar(val):
    return Multivector({(): val})

def e(i):
    return Multivector({(i,): 1.0})

# ========================
# Example usage:
# ========================
if __name__ == "__main__":
    # Define basis vectors for 3D Euclidean space.
    e1 = e(1)
    e2 = e(2)
    e3 = e(3)

    # Compute the pseudoscalar I = e1 * e2 * e3.
    I = e1 * e2 * e3

    # Define a sample Hamiltonian as a bivector; for example, H = e1
    ↪  * e2.
    H = e1 * e2

    # Define an initial quantum state  as a multivector.
    # Example:  = 1 + 0.5e1.
    Psi = scalar(1.0) + 0.5 * e1

    hbar = 1.0  # Planck's constant (in natural units)
    t = 0.1     # time interval

    # Compute the time evolution operator: U(t) = exp(- (I * H * t)
    ↪  / )
    U = time_evolution_operator(H, I, t, hbar)

    # Evolve the quantum state:  (t) = U(t) *
    Psi_t = U * Psi

    # Compute the time derivative d/dt at t = 0.
    dPsi_dt = time_derivative(Psi, H, I, hbar)

    print("Initial quantum state :")
    print(Psi)
    print("\nHamiltonian H (as a bivector):")
    print(H)
    print("\nPseudoscalar I:")
```

```python
print(I)
print("\nTime evolution operator U(t):")
print(U)
print("\nEvolved quantum state (t):")
print(Psi_t)
print("\nTime derivative d/dt:")
print(dPsi_dt)
```

Chapter 34

Wave Functions as Multivectors: Representation and Computation

Multivector Formulation of Quantum Wave Functions

In conventional quantum mechanics a wave function $\psi(x)$ is defined as a complex-valued function over the spatial domain. In the geometric algebra formulation the same physical information is embedded in a multivector field. Instead of representing state amplitudes exclusively by complex numbers, the wave function is expanded in a basis of geometric elements, for example,

$$\Psi(x) = \psi_0(x) + \psi_1(x)e_1 + \psi_{12}(x)e_1 e_2 + \psi_{123}(x)e_1 e_2 e_3 + \cdots,$$

where each component $\psi_k(x)$ is a real-valued function. In this representation the scalar part carries the conventional amplitude, while the higher-grade terms encode phase, geometric orientation, and intrinsic symmetry information. The multivector structure naturally incorporates a rotation mechanism that substitutes the conventional imaginary unit by specific bivectors or pseudoscalars, thereby providing an invariant framework under geometrical transformations.

Transformation and Invariance Properties

The geometric product used in the multivector formalism preserves both metric and orientation information. Under spatial rotations and reflections, the individual components of $\Psi(x)$ transform in a coordinated manner such that the overall multivector retains its physical meaning. If R is a rotor representing a rotation then the transformed state is given by

$$\Psi'(x) = R\Psi(x)R^{-1}.$$

This operation maintains the integrity of the wave function's geometric structure. Invariance under coordinate transformations is achieved without resorting to extraneous complex conjugation procedures, as the necessary phase rotations are generated intrinsically by the bivector parts of the multivector. Such a formulation leads to a more robust handling of symmetries and offers a direct connection between symmetry generators in the algebra and conserved quantities in the physical system.

Computational Efficiency and Benefits

Reinterpreting wave functions as multivectors offers significant computational advantages in numerical simulations of quantum systems. The unified representation consolidates scalar and vector operations into a single algebraic product. This consolidation reduces the complexity inherent in managing separate number fields and vector spaces. Moreover, the multivector formulation leads to more compact algorithms by exploiting the distributive and associative properties of the geometric product. As operations such as differentiation and integration of wave functions are carried out, the algebraic structure ensures that metric invariance is maintained, avoiding the pitfalls of coordinate singularities.

The ability to represent both amplitude and phase within one algebraic object enables more natural discretization schemes when simulating partial differential equations that govern quantum dynamics. Data structures designed for multivector components can efficiently store sparse representations, which is beneficial for high-performance computing applications in quantum mechanics. Rigorous error control is also facilitated by the coherent structure of the multivector framework.

Representative Implementation Function

A representative function for converting a complex amplitude to its corresponding multivector form encapsulates the core idea behind the wave function reinterpretation. In the geometric algebra approach the standard mapping

$$\Psi = \Re(\psi) + \Im(\psi)\, I$$

associates the real part of the complex wave function with the scalar grade and the imaginary part with a predefined bivector element I. The following Python function illustrates this concept. It takes a complex amplitude as input and returns a multivector represented as a dictionary. Here the scalar component is stored with the key (), while the bivector corresponding to the imaginary unit is stored with the key $(1, 2)$.

```
def wave_function_to_multivector(psi):
    """
    Convert a complex quantum wave function amplitude into a
    ↪ multivector representation.

    The mapping follows:
        Psi = Re(psi) + Im(psi)*I,
    where the scalar part is recorded under the empty tuple () and
    ↪ the bivector
    corresponding to the imaginary unit (I) is represented by the
    ↪ basis (1, 2).

    Parameters:
        psi (complex): A complex number representing the quantum
        ↪ amplitude.

    Returns:
        dict: A dictionary representing the multivector with keys as
        ↪ basis blades.
            For example, {(): Re(psi), (1, 2): Im(psi)}.
    """
    return {(): psi.real, (1, 2): psi.imag}
```

Python Code Snippet

```
import math

def wave_function_to_multivector(psi):
    """
```

```
    Convert a complex quantum wave function amplitude into a
    ↪   multivector representation.

    The mapping follows:
        = Re(psi) + Im(psi)*I,
    where:
      - The scalar part is stored with the key ().
      - The imaginary unit is represented by a predefined bivector
    ↪   with key (1,2),
        taking advantage of the fact that in a 2D Euclidean Clifford
    ↪   Algebra,
        (1,2)**2 = -1.

    Parameters:
        psi (complex): A complex number representing the quantum
    ↪   amplitude.

    Returns:
        dict: A dictionary representing the multivector. For
    ↪   example, {(): Re(psi), (1, 2): Im(psi)}.
    """
    return {(): psi.real, (1, 2): psi.imag}

def multiply_blades(b1, b2):
    """
    Compute the geometric product of two basis blades.

    Both b1 and b2 are represented as tuples of sorted basis
    ↪   indices.
    The product is obtained by concatenating the lists and then
    ↪   performing bubble-sort
    to reorder the basis vectors to canonical order (ascending
    ↪   order) while counting the sign changes.
    Duplicate basis vectors are removed in pairs (since in an
    ↪   orthonormal Euclidean metric, e_i^2 = 1),
    effectively performing the cancellation.

    Parameters:
        b1 (tuple): Basis blade as a tuple, e.g., () for a scalar,
    ↪   (1,) for e1, (1,2) for e1e2.
        b2 (tuple): Another basis blade.

    Returns:
        tuple: A pair (s, blade) where s is +1 or -1 (the sign
    ↪   factor from reordering)
            and blade is the resulting basis blade as a sorted
    ↪   tuple.
    """
    # Concatenate the lists of indices.
    combined = list(b1) + list(b2)
    sign = 1
    # Bubble sort to reorder combined list to canonical ascending
    ↪   order;
```

```
    # count each adjacent swap as a sign change.
    n = len(combined)
    for i in range(n):
        for j in range(n-1, i, -1):
            if combined[j-1] > combined[j]:
                combined[j-1], combined[j] = combined[j],
                ↪ combined[j-1]
                sign = -sign
    # Remove duplicates in pairs (each duplicate pair contributes
    ↪ e_i^2 = 1).
    i = 0
    result = []
    while i < len(combined):
        if i < len(combined)-1 and combined[i] == combined[i+1]:
            i += 2  # cancel the pair
        else:
            result.append(combined[i])
            i += 1
    return sign, tuple(result)

def geometric_product(M, N):
    """
    Compute the geometric product of two multivectors M and N.

    Multivectors are represented as dictionaries mapping basis
    ↪ blades (tuples) to coefficients.
    The geometric product is defined by bilinearity:
        (sum_a A_a e_a) (sum_b B_b e_b) = sum_a,b (A_a B_b) (e_a
    ↪ e_b),
    where the product of basis blades is determined by the
    ↪ multiply_blades function.

    Parameters:
        M (dict): First multivector, e.g., {(): 1.0, (1, 2): 2.0}
        N (dict): Second multivector.

    Returns:
        dict: The resulting multivector after performing the
        ↪ geometric product.
    """
    result = {}
    for blade1, coeff1 in M.items():
        for blade2, coeff2 in N.items():
            s, new_blade = multiply_blades(blade1, blade2)
            result[new_blade] = result.get(new_blade, 0) + coeff1 *
            ↪ coeff2 * s
    return result

def reverse(M):
    """
    Compute the reversion (or reversal) of a multivector.
```

```
    For each basis blade of grade r (length of the key tuple), the
    ↪ reversal is defined as:
        reverse(e_{i_1}e_{i_2}...e_{i_r}) = (-1)^(r*(r-1)/2)
        ↪ e_{i_1}e_{i_2}...e_{i_r}.

    This operation is particularly important because for a
    ↪ normalized rotor R,
    the inverse of R is given by its reverse.

    Parameters:
        M (dict): A multivector represented as a dictionary.

    Returns:
        dict: The reversed multivector.
    """
    result = {}
    for blade, coeff in M.items():
        r = len(blade)
        factor = (-1) ** (r * (r - 1) // 2)
        result[blade] = coeff * factor
    return result

def rotor_transform(psi, R):
    """
    Apply a rotor transformation to a multivector (e.g., a quantum
    ↪ wave function).

    The transformation is given by:
        psi' = R * psi * R^{-1},
    and for a unit rotor R (i.e., R * reverse(R) = 1), the inverse
    ↪ is simply the reversion.

    Parameters:
        psi (dict): The multivector representation of the state.
        R (dict): The rotor multivector.

    Returns:
        dict: The transformed multivector.
    """
    return geometric_product(geometric_product(R, psi), reverse(R))

def format_multivector(M):
    """
    Generate a formatted string for a multivector for display.

    Scalars are displayed normally, while non-scalar parts are shown
    ↪ as a coefficient
    multiplied by a basis string (e.g., "e1e2" for the basis blade
    ↪ (1,2)).

    Parameters:
        M (dict): A multivector.
```

```
    Returns:
        str: A human-readable string representation.
    """
    terms = []
    for blade, coeff in sorted(M.items(), key=lambda x: (len(x[0]),
    ↪  x[0])):
        if abs(coeff) < 1e-12:
            continue
        if blade == ():
            term = f"{coeff:.3g}"
        else:
            basis = "".join([f"e{n}" for n in blade])
            term = f"{coeff:.3g}{basis}"
        terms.append(term)
    return " + ".join(terms) if terms else "0"

# -------------------- Example Usage --------------------

if __name__ == "__main__":
    # Example: Convert a complex quantum wave function amplitude
    ↪  into a multivector.
    # Suppose the complex amplitude is psi = 3 + 4j.
    psi_complex = 3 + 4j
    Psi_mv = wave_function_to_multivector(psi_complex)
    print("Wave function multivector representation:")
    print(format_multivector(Psi_mv))

    # Define a rotor representing a rotation by theta about the
    ↪  plane defined by e1 and e2.
    # The rotor has the form: R = cos(theta/2) - I*sin(theta/2),
    ↪  where I is represented by (1,2).
    theta = math.pi / 4   # 45 degrees rotation
    R = {(): math.cos(theta/2), (1,2): -math.sin(theta/2)}  # Note
    ↪  the minus sign before sin(theta/2)

    # Apply the rotor transformation: ' = R  R^{-1} (with R^{-1} =
    ↪  reverse(R) for a unit rotor).
    Psi_rotated = rotor_transform(Psi_mv, R)
    print("\nRotated wave function multivector:")
    print(format_multivector(Psi_rotated))

    # Demonstrate geometric product on two simple multivectors.
    # Let A = 2 + 3I and B = 1 - I, where I = (1,2).
    A = {(): 2, (1,2): 3}
    B = {(): 1, (1,2): -1}
    product = geometric_product(A, B)
    print("\nGeometric product of A and B:")
    print(format_multivector(product))
```

Chapter 35

Operator Theory in GA: Linear and Nonlinear Aspects

Operator Formalism in Geometric Algebra

Operator theory within geometric algebra is based on establishing mappings on the space of multivectors that reflect both algebraic and geometric properties. An operator $\mathcal{O}$ is defined by its action on individual basis blades, extended to the entire algebra by linearity. For a multivector M, decomposed as

$$M = \sum_I m_I \, e_I,$$

the operator acts according to

$$\mathcal{O}(M) = \sum_I m_I \, \mathcal{O}(e_I),$$

where each basis element e_I carries specific geometric meaning. The formalism subsumes traditional differential operators as well as transformations that preserve or modify the multivector structure. This construction provides a natural framework for characterizing rotations, reflections, and other symmetry transformations in a unified algebraic language.

Linear Operator Methods

Linear operators $\mathcal{L}$ in geometric algebra satisfy the property

$$\mathcal{L}(A+B) = \mathcal{L}(A) + \mathcal{L}(B), \quad \mathcal{L}(cA) = c\,\mathcal{L}(A),$$

for any multivectors A, B and scalar c. Such operators may arise from projection operations, rotations, or differential mappings, and are closely tied to the structure of the underlying vector space. The representation of a linear operator in GA is often encoded in terms of its action on a predefined basis. In computational environments, one strategy is to represent both the operator and the multivector as dictionaries with basis blades as keys and corresponding coefficients as values.

A representative Python function that captures the essence of applying a linear operator to a multivector is provided below. The operator is modeled as a mapping from each blade to a sub-dictionary that encodes the transformation on that blade. The function iterates over the input multivector and accumulates the contributions of the linear transformation.

```
def apply_linear_operator(operator, multivector):
    """
    Apply a linear operator to a multivector.

    The operator is represented as a dictionary that maps each basis
        blade to a sub-dictionary,
    where the keys of the sub-dictionary represent the basis blades
        of the output and the values
    are the corresponding transformation factors. The multivector is
        represented as a dictionary
    with basis blades as keys and their coefficients as values.

    Parameters:
        operator (dict): A mapping from basis blades (tuples) to
            dictionaries that define the linear
                        transformation for each blade.
        multivector (dict): A dictionary representing the
            multivector to be transformed.

    Returns:
        dict: A new multivector resulting from the application of
            the linear operator.
    """
    result = {}
    for blade, coeff in multivector.items():
        if blade in operator:
            transformation = operator[blade]
```

```
        for new_blade, factor in transformation.items():
            result[new_blade] = result.get(new_blade, 0) + coeff
            ↪    * factor
    else:
        result[blade] = result.get(blade, 0) + coeff
return result
```

The computational structure of the function exploits the distributivity and linearity inherent to the operator, ensuring that the transformation aligns with the geometric product and overall GA framework. Such implementations are fundamental to high-performance simulations where operator actions are applied repeatedly to complex multivector states.

Nonlinear Operator Techniques

In contrast to linear operators, certain physical systems require the application of operators that exhibit nonlinear characteristics. Nonlinear operators in GA are defined by their departure from the superposition principle. Their action is often described by polynomial or analytic functions of the multivector arguments. For instance, a nonlinear operator $\mathcal{N}$ may be constructed as a power series in the geometric product:

$$\mathcal{N}(M) = \sum_{k=0}^{\infty} a_k \, M^k,$$

with coefficients a_k selected to capture the specific nonlinear dynamics of the system.

The algebraic structure of geometric algebra facilitates the treatment of such nonlinear operators by providing a natural means to encode multivector products and cancellations among basis elements. The ability to manipulate nonlinear combinations of blades explicitly aids in modeling complex interactions, such as those found in nonlinear wave propagation or chaotic dynamical systems. This theoretical framework allows for a systematic study of stability properties and the emergence of invariants even in the presence of nonlinearities.

Operator Commutation and Algebraic Structures

A core aspect of operator theory is the analysis of commutation relations. For two operators $\mathcal{A}$ and $\mathcal{B}$ defined over a GA space, the commutator is given by

$$[\mathcal{A}, \mathcal{B}] = \mathcal{A}\mathcal{B} - \mathcal{B}\mathcal{A}.$$

This expression captures the degree to which the actions of $\mathcal{A}$ and $\mathcal{B}$ are intertwined. In many physical applications, particularly within quantum mechanics and field theory, such commutation relations are directly linked to conserved quantities and symmetry transformations.

The noncommutative properties of the geometric product grant additional depth to the analysis. Operators that do not commute may generate algebraic structures analogous to Lie algebras, wherein the commutators impose constraints on the evolution and interaction of physical observables. The study of these algebraic relations is essential for understanding the spectral properties of operators and for the development of perturbation methods in both linear and nonlinear regimes.

The synthesis of operator theory with geometric algebra establishes a robust computational framework. The intrinsic interplay of linear and nonlinear operators, together with their commutative or noncommutative characteristics, enables the concise expression of complex physical phenomena and supports the development of efficient algorithms for simulation and analysis.

Python Code Snippet

```
# Python Code for Operator Theory in Geometric Algebra
# This code snippet demonstrates several key algorithms:
# 1. Applying a linear operator to a multivector.
# 2. Computing the geometric product of multivectors.
# 3. Implementing a nonlinear operator as a power series expansion.
# 4. Calculating the commutator of two operators.

def add_mv(mv1, mv2):
    """
    Add two multivectors represented as dictionaries.
    Each key is a basis blade (tuple) and each value is its
    ↪ coefficient.
```

```python
    """
    result = mv1.copy()
    for blade, coeff in mv2.items():
        if blade in result:
            result[blade] += coeff
        else:
            result[blade] = coeff
    return result

def subtract_mv(mv1, mv2):
    """
    Subtract multivector mv2 from mv1.
    """
    result = mv1.copy()
    for blade, coeff in mv2.items():
        if blade in result:
            result[blade] -= coeff
        else:
            result[blade] = -coeff
    return result

def scale_mv(mv, scalar):
    """
    Scale a multivector by a scalar.
    """
    return {blade: coeff * scalar for blade, coeff in mv.items()}

def bubble_sort_sign(lst):
    """
    Perform bubble sort on the list while tracking the sign change
    ↪  due to swaps.
    Returns the sorted list and the overall sign.
    """
    L = lst.copy()
    sign = 1
    n = len(L)
    for i in range(n):
        for j in range(0, n - i - 1):
            if L[j] > L[j + 1]:
                L[j], L[j+1] = L[j+1], L[j]
                sign *= -1
    return L, sign

def multiply_blades(blade1, blade2):
    """
    Multiply two basis blades.

    Each blade is represented as a tuple of basis vector labels
    ↪  (e.g., ('e1', 'e2')).
    The geometric product is computed by:
        1. Concatenating the two blades.
        2. Sorting the combined list with sign changes from each swap.
```

    3. *Canceling duplicate basis vectors (using e_i * e_i = 1 for*
    ↪ *Euclidean signature).*

    Returns a tuple (result_blade, sign) where result_blade is the
    ↪ *canonical tuple.*
    """
    # Concatenate the basis vectors from both blades.
    combined = list(blade1) + list(blade2)
    # Sort the combined list and get the sign change.
    sorted_combined, sign = bubble_sort_sign(combined)
    # Cancel duplicate basis vectors.
    result_blade = []
    i = 0
    n = len(sorted_combined)
    while i < n:
        count = 1
        # Count duplicates.
        while i + 1 < n and sorted_combined[i] == sorted_combined[i
        ↪ + 1]:
            count += 1
            i += 1
        # If count is odd, one copy remains; if even, they cancel
        ↪ out.
        if count % 2 == 1:
            result_blade.append(sorted_combined[i])
        i += 1
    return (tuple(result_blade), sign)

def geometric_product(mv1, mv2):
    """
    Compute the geometric product of two multivectors.

    A multivector is represented as a dictionary with keys as basis
    ↪ *blades (tuples)*
    and values as their coefficients.
    """
    result = {}
    for blade1, coeff1 in mv1.items():
        for blade2, coeff2 in mv2.items():
            new_blade, sign = multiply_blades(blade1, blade2)
            new_coeff = coeff1 * coeff2 * sign
            if new_blade in result:
                result[new_blade] += new_coeff
            else:
                result[new_blade] = new_coeff
    return result

def multivector_power(mv, k):
    """
    Compute the power of a multivector mv raised to exponent k.

    The scalar 1 is represented as {(): 1}, where () denotes the
    ↪ *scalar (grade 0).*

```
"""
if k == 0:
    return {(): 1}
result = mv
for _ in range(1, k):
    result = geometric_product(result, mv)
return result

def apply_linear_operator(operator, mv):
    """
    Apply a linear operator to a multivector.

    The operator is represented as a dictionary mapping each basis
     ↪  blade to a sub-dictionary.
    The sub-dictionary defines how that blade is transformed into
     ↪  output blades with factors.
    The multivector is a dictionary with basis blades as keys and
     ↪  their coefficients as values.
    """
    result = {}
    for blade, coeff in mv.items():
        if blade in operator:
            transformation = operator[blade]
            for new_blade, factor in transformation.items():
                result[new_blade] = result.get(new_blade, 0) + coeff
                 ↪  * factor
        else:
            result[blade] = result.get(blade, 0) + coeff
    return result

def apply_nonlinear_operator(nonlinear_coeffs, mv, max_power):
    """
    Apply a nonlinear operator to a multivector using a power series
     ↪  expansion.

    The nonlinear operator is defined as:
        N(mv) = ^(max_power) a_(mv) ^k
    where nonlinear_coeffs is a dictionary mapping exponent k to
     ↪  coefficient a.
    """
    result = {}
    for k in range(max_power + 1):
        # Compute mv^k; note that for k=0, we use the scalar 1.
        term = multivector_power(mv, k)
        a_k = nonlinear_coeffs.get(k, 0)
        term_scaled = scale_mv(term, a_k)
        result = add_mv(result, term_scaled)
    return result

def commutator(operator_A, operator_B, mv):
    """
    Compute the commutator [A, B] on a multivector mv using linear
     ↪  operators.
```

```
    The commutator is defined as:
        [A, B](mv) = A(B(mv)) - B(A(mv))
    Both operators are represented as dictionaries.
    """
    A_then_B = apply_linear_operator(operator_A,
    ↪    apply_linear_operator(operator_B, mv))
    B_then_A = apply_linear_operator(operator_B,
    ↪    apply_linear_operator(operator_A, mv))
    return subtract_mv(A_then_B, B_then_A)

# Example usage:

# Define an example multivector: M = 2*e1 + 3*e2
M = {('e1',): 2, ('e2',): 3}

# Define a linear operator.
# For basis blade ('e1',), it transforms to 0.5*('e1',) and
↪    1.5*('e2',);
# For ('e2',), it leaves the component unchanged.
linear_op = {
    ('e1',): {('e1',): 0.5, ('e2',): 1.5},
    ('e2',): {('e2',): 1.0}
}

# Apply the linear operator to M.
M_transformed = apply_linear_operator(linear_op, M)
print("Result of applying linear operator to M:")
print(M_transformed)

# Define nonlinear operator coefficients for a power series:
# N(M) = 1 + 0.5*M + 0.2*M^2.
nonlinear_coeffs = {0: 1, 1: 0.5, 2: 0.2}
M_nonlinear = apply_nonlinear_operator(nonlinear_coeffs, M,
↪    max_power=2)
print("\nResult of applying nonlinear operator to M:")
print(M_nonlinear)

# Define a second linear operator for illustration.
# This operator doubles the ('e1',) component and negates the
↪    ('e2',) component.
linear_op_B = {
    ('e1',): {('e1',): 2},
    ('e2',): {('e2',): -1}
}

# Compute the commutator of linear_op and linear_op_B applied to M.
comm_result = commutator(linear_op, linear_op_B, M)
print("\nCommutator [linear_op, linear_op_B] applied to M:")
print(comm_result)
```

Chapter 36

Angular Momentum in GA: Computational Representations

Mathematical Foundations

Within the framework of Geometric Algebra (GA), angular momentum attains an elegant representation in the form of bivectors. In a three-dimensional Euclidean setting, the classical notion of angular momentum—traditionally expressed as a cross product—is generalized by defining it as the wedge product of the position vector r and the momentum vector p. This yields the bivector

$$L = r \wedge p,$$

which encapsulates both the magnitude and the oriented plane of rotation. The antisymmetric property of the wedge product, paired with the intrinsic structure of blades in GA, ensures that the bivector formalism not only aligns with classical mechanics but also extends naturally to more complex spatial transformations.

Computational Representation of Angular Momentum

The computational realization of angular momentum in GA typically involves representing multivectors as data structures, such as dictionaries, that map basis blades to their corresponding scalar coefficients. In this setting, the vectors r and p are encoded as collections where each key denotes a one-dimensional blade (e.g., ('e1',) denotes the x-axis basis element). The decisive operation is the wedge product, which, due to its antisymmetric nature, demands careful ordering of the basis elements to correctly account for sign changes and cancellations.

An illustrative Python function demonstrates the method for calculating the angular momentum bivector. The function iterates over all pairs of basis components from the position and momentum vectors. For each pair, if the basis elements are distinct, the union of the bases is ordered canonically, and an appropriate sign (representing the parity of the permutation) is applied. The result is accumulated into a dictionary that represents the bivector corresponding to the angular momentum.

```
def compute_angular_momentum(r, p):
    """
    Compute the angular momentum bivector given position and
    ↪ momentum vectors.

    Position vector r and momentum vector p are represented as
    ↪ dictionaries, where keys are tuples
    representing individual basis blades (e.g., ('e1',)) and values
    ↪ are the associated scalar components.

    The angular momentum is defined as the wedge product: L = r  p.
    In three-dimensional space, the wedge product yields a bivector
    ↪ that encapsulates the oriented plane
    and magnitude of the angular momentum.

    Returns:
        A dictionary representing the angular momentum bivector.
    """
    L = {}
    for blade_r, coeff_r in r.items():
        for blade_p, coeff_p in p.items():
            if blade_r == blade_p:
                continue
            combined = blade_r + blade_p
            sorted_basis = tuple(sorted(combined))
            sign = 1  # Determination of sign based on permutation
            ↪ parity omitted for brevity.
            L[sorted_basis] = L.get(sorted_basis, 0) + sign *
            ↪ coeff_r * coeff_p
    return L
```

Algorithmic Considerations and Analysis

The efficiency and accuracy of angular momentum computations in GA are deeply contingent on the meticulous management of the wedge product's antisymmetry. Given that the operation fundamentally relies on the canonical ordering of basis blades, the algorithm must recognize and cancel contributions from repeated elements, thereby ensuring that only the unique bivector components are preserved.

Data structures representing multivectors facilitate rapid look-up and accumulation of coefficients, and their use is central to maintaining computational efficiency. The algorithm systematically iterates over the component pairs of the input vectors, applying a permutation routine to order the elements and determine the associated sign change. Although the detailed calculation of permutation parity is not explicitly shown in the function, its placeholder underscores the importance of this step in preserving the mathematical rigor of the method.

By encoding angular momentum as a bivector, GA seamlessly bridges the geometric intuition with algebraic computational techniques. This representation serves as a powerful tool in both symbolic manipulation and numerical simulation, allowing for the precise evaluation of rotational dynamics. The strategic combination of mathematical insight with computational methodologies fosters an environment in which complex physical phenomena can be modeled with both clarity and efficiency.

Python Code Snippet

```
def permutation_parity(seq):
    """
    Compute the parity (sign) of the permutation needed to sort the
    ↪ sequence.

    This function counts the number of inversions in the sequence.
    Returns:
        1 if there are an even number of inversions, -1 if odd.
    """
    inversions = 0
    for i in range(len(seq)):
        for j in range(i + 1, len(seq)):
            if seq[i] > seq[j]:
                inversions += 1
    return -1 if inversions % 2 else 1
```

```python
def wedge_basis(blade1, blade2):
    """
    Compute the wedge product of two basis blades.

    Each blade is represented as a tuple of basis elements (e.g.,
    ↪ ('e1',)).
    If the blades share any common basis element, the wedge product
    ↪ is zero,
    as the antisymmetry of the wedge product annihilates repeated
    ↪ elements.

    Args:
        blade1 (tuple): The first basis blade.
        blade2 (tuple): The second basis blade.

    Returns:
        tuple: A sorted tuple representing the new basis blade.
        int: The sign factor (±1) resulting from the permutation
        ↪ required to
            sort the combined basis. Returns 0 if the wedge is
            ↪ zero.
    """
    # Check for repeated basis elements (which would cancel out).
    if set(blade1) & set(blade2):
        return (), 0
    combined = blade1 + blade2
    # Compute the sign arising from the permutation that sorts the
    ↪ combined basis.
    sign = permutation_parity(list(combined))
    sorted_basis = tuple(sorted(combined))
    return sorted_basis, sign

def wedge_product(mv1, mv2):
    """
    Compute the wedge product between two multivectors.

    A multivector is represented as a dictionary where keys are
    ↪ tuples corresponding
    to basis blades (e.g., ('e1',) for a vector) and values are the
    ↪ scalar components.

    Args:
        mv1 (dict): The first multivector.
        mv2 (dict): The second multivector.

    Returns:
        dict: A dictionary representing the resultant multivector
        ↪ from the wedge product.
    """
    result = {}
    # Iterate over all pairs of basis blades from mv1 and mv2.
    for blade1, coeff1 in mv1.items():
```

```python
        for blade2, coeff2 in mv2.items():
            new_blade, sign = wedge_basis(blade1, blade2)
            # If sign is zero, the wedge product of these blades is
            ↪ zero.
            if sign == 0:
                continue
            # Accumulate the result in the dictionary.
            result[new_blade] = result.get(new_blade, 0) + sign *
            ↪ coeff1 * coeff2
    return result

def compute_angular_momentum(r, p):
    """
    Compute the angular momentum bivector given position and
    ↪ momentum vectors.

    In Geometric Algebra, the angular momentum is defined as the
    ↪ wedge product:
        L = r  p
    where r and p are vectors represented as multivectors
    ↪ (dictionaries).

    Args:
        r (dict): The position vector with keys as one-dimensional
        ↪ basis blades.
        p (dict): The momentum vector with keys as one-dimensional
        ↪ basis blades.

    Returns:
        dict: A dictionary representing the angular momentum
        ↪ bivector.
    """
    return wedge_product(r, p)

# Example usage of the above functions.
if __name__ == "__main__":
    # Define a position vector r and a momentum vector p in
    ↪ three-dimensional space.
    # Each vector is represented as a dictionary where the key is a
    ↪ tuple signifying the basis.
    r = {('e1',): 1.0, ('e2',): 2.0, ('e3',): 3.0}   # Example
    ↪ position vector components.
    p = {('e1',): 4.0, ('e2',): 5.0, ('e3',): 6.0}   # Example
    ↪ momentum vector components.

    # Compute the angular momentum bivector using the wedge product
    ↪ (L = r  p).
    L = compute_angular_momentum(r, p)

    # Display the computed angular momentum bivector.
    print("Angular Momentum (bivector representation):")
    for blade, coeff in L.items():
```

```
print(f"{blade}: {coeff}")
```

Chapter 37

Pauli Algebra: Linking Spin and Geometry

Algebraic Structure and Mapping in Geometric Algebra

In the framework of geometric algebra, the well-known Pauli matrices from quantum mechanics find an elegant reinterpretation. The correspondence is established by associating the three canonical Pauli matrices, denoted by σ_1, σ_2, and σ_3, with the unit basis vectors e_1, e_2, and e_3 in a three-dimensional Euclidean space. The geometric product of these basis vectors decomposes naturally into a scalar (symmetric) part and a bivector (antisymmetric) part. In particular, for two orthogonal unit vectors the geometric product reduces to the wedge product:

$$e_i e_j = e_i \cdot e_j + e_i \wedge e_j = e_i \wedge e_j, \quad i \neq j.$$

This relation mirrors the algebra of the Pauli matrices, wherein the product

$$\sigma_i \sigma_j = \delta_{ij} + i\, \epsilon_{ijk} \sigma_k$$

splits into a contribution that enforces orthonormality and a term encapsulating the intrinsic rotation, with the imaginary unit i acquiring a geometric significance when interpreted as a bivector. By replacing the abstract complex unit with a well-defined bivector, geometric algebra unifies the language of rotations and spin. The bivector $e_i \wedge e_j$ represents the oriented plane spanned by the two basis vectors, and its inherent antisymmetry encodes the structure of spin interactions.

Spinor Representation and Rotor Construction

Spinors in geometric algebra are constructed as even-grade multivectors that encode rotations via the exponential of bivectors. A rotation by an angle θ in the plane defined by a bivector B is effected by the rotor

$$R = \cos\left(\frac{\theta}{2}\right) - \hat{B}\sin\left(\frac{\theta}{2}\right),$$

where $\hat{B}$ is the normalized bivector, satisfying $\hat{B}^2 = -1$. This formulation replaces the standard complex exponential with a purely geometric one and closely aligns with the operations expressed by the Pauli matrices when modeling spin dynamics. Computation of such rotors is central to simulations of spinor dynamics and quantum transformations, as it enables the direct manipulation of the state without resorting to matrix representations.

A computational implementation typically represents multivectors as dictionaries mapping basis blade identifiers to their scalar coefficients. Efficient algorithms require the evaluation of functions such as the rotor exponential; for instance, the following Python function computes the rotor corresponding to a given bivector and rotation angle.

```
def compute_rotor(bivector, theta):
    """
    Compute the rotor from a given bivector to model a spin
    ↪ rotation.

    Given a pure bivector 'B' represented as a dictionary mapping a
    ↪ tuple of basis elements
    (e.g., ('e1','e2')) to its scalar coefficient, and a rotation
    ↪ angle 'theta' (in radians),
    the rotor is defined as:
        R = cos(theta/2) - (B/|B|)*sin(theta/2)
    where |B| is the magnitude of the bivector. The normalization of
    ↪ B ensures that the
    bivector acts analogously to an imaginary unit with B^2 =
    ↪ -|B|^2.
```

```
The rotor 'R' is returned as a multivector in dictionary form,
↪ containing both the scalar
part and the bivector part.

Args:
    bivector (dict): A dictionary representing a pure bivector
    ↪ with keys as tuples of
                basis elements (e.g., ('e1','e2')) and
                    ↪ values as the scalar coefficients.
    theta (float): The rotation angle in radians.

Returns:
    dict: A multivector representing the rotor.
"""
# Compute the magnitude of the bivector.
magnitude = sum(coefficient**2 for coefficient in
↪ bivector.values())**0.5
if magnitude == 0:
    raise ValueError("Zero magnitude bivector cannot be
    ↪ normalized.")

# Compute the normalized bivector.
normalized_bivector = {blade: coeff / magnitude for blade, coeff
↪ in bivector.items()}

# Calculate the scalar and bivector parts of the rotor.
scalar_part = cos(theta / 2)
bivector_part = -sin(theta / 2)

# Assemble the rotor as a multivector.
rotor = {(): scalar_part}  # The empty tuple represents the
↪ scalar component.
for blade, coeff in normalized_bivector.items():
    rotor[blade] = bivector_part * coeff

return rotor
```

Computational Considerations in Pauli-Algebraic Operations

The computational modeling of spin systems via Pauli algebra within the geometric algebra framework demands careful handling of algebraic structures and data representations. By encoding elements as dictionaries, the algorithms allow straightforward implementation of the geometric (and wedge) products with inherent antisymmetry. The structure of the rotor as an even-grade multivector underpins efficient simulation of spin transformations.

Such methods facilitate the direct translation of the algebraic relations between Pauli matrices into computational routines. The normalization process, the extraction of the magnitude for bivectors, and the precise composition of the rotor are implemented in a self-contained manner. This approach not only preserves the mathematical rigor of the underlying physical models but also supports numerical stability and efficiency in simulations of quantum spin dynamics.

The detailed mapping from the abstract Pauli matrices to geometric algebra elements, along with the computational routines for rotor formation, exemplifies how spin, a fundamentally quantum phenomenon, is modeled using techniques from computational physics. This synthesis of algebraic insight and numerical methods provides a versatile toolkit for simulating systems where spin and geometry are intimately linked.

Python Code Snippet

```python
import math

def compute_magnitude(bivector):
    """
    Compute the magnitude (norm) of a pure bivector represented as a
      dictionary.
    The magnitude is defined as the square root of the sum of the
      squares of the coefficients.
    """
    return math.sqrt(sum(coefficient**2 for coefficient in
      bivector.values()))

def normalize_bivector(bivector):
    """
    Normalize a pure bivector so that its magnitude becomes 1.

    Args:
        bivector (dict): A dictionary representing a pure bivector
          where keys are tuples
                         defining the basis blade (e.g., ('e1',
          'e2')) and values are the scalar
                         coefficients.

    Returns:
        dict: A normalized bivector.
    """
    magnitude = compute_magnitude(bivector)
    if magnitude == 0:
```

```python
        raise ValueError("Cannot normalize a bivector with zero
        ↪   magnitude.")
    return {blade: coeff / magnitude for blade, coeff in
        ↪   bivector.items()}

def compute_rotor(bivector, theta):
    """
    Compute the rotor from a given bivector to model a spin
    ↪   rotation.

    In the geometric algebra framework, a rotation by an angle theta
    ↪   in the plane defined
    by a bivector B is effected by the rotor:
        R = cos(theta/2) - (B/|B|) * sin(theta/2)
    where |B| is the magnitude of B. The bivector B is assumed to be
    ↪   a pure bivector and, once
    normalized, acts analogously to the imaginary unit with B^2 =
    ↪   -1.

    Args:
        bivector (dict): A dictionary representing a pure bivector
        ↪   (e.g., {('e1','e2'): 1.0}).
        theta (float): The rotation angle in radians.

    Returns:
        dict: A multivector representing the rotor, with the scalar
        ↪   part under key ().
    """
    # Normalize the bivector to ensure it behaves like an imaginary
    ↪   unit.
    normalized_bivector = normalize_bivector(bivector)

    # Compute the scalar part and the factor for the bivector part
    ↪   of the rotor.
    scalar_part = math.cos(theta / 2)
    bivector_part_coeff = -math.sin(theta / 2)

    # Assemble the rotor as a multivector.
    # The scalar component is stored with the empty tuple as key.
    rotor = {(): scalar_part}
    for blade, coeff in normalized_bivector.items():
        rotor[blade] = bivector_part_coeff * coeff

    return rotor

def multivector_to_string(multivector):
    """
    Convert a multivector (represented as a dictionary) into a
    ↪   human-readable string.

    Args:
        multivector (dict): The multivector with keys as basis blade
        ↪   tuples and values as coefficients.
```

```
    Returns:
        str: A string representation of the multivector.
    """
    terms = []
    # Sort keys by grade and lexicographical order for consistency.
    for blade in sorted(multivector.keys(), key=lambda x: (len(x),
    ↪    x)):
        coeff = multivector[blade]
        # Format the scalar part.
        if blade == ():
            term = f"{coeff:.4f}"
        else:
            # Concatenate the basis elements (e.g., ('e1','e2') ->
            ↪    "e1e2").
            blade_str = "".join(blade)
            term = f"{coeff:.4f}{blade_str}"
        terms.append(term)
    return " + ".join(terms)

# Demonstration of rotor computation for spin rotation.
if __name__ == "__main__":
    # Example 1: Compute the rotor for a bivector in the e1^e2
    ↪    plane.
    bivector1 = {('e1', 'e2'): 1.0}  # A pure bivector representing
    ↪    the plane spanned by e1 and e2.
    theta1 = math.pi / 2  # 90 degrees rotation.

    rotor1 = compute_rotor(bivector1, theta1)
    print("Rotor for bivector {('e1','e2'): 1.0} with 90°
    ↪    rotation:")
    print(multivector_to_string(rotor1))

    # Example 2: Compute the rotor for a bivector in the e2^e3 plane
    ↪    with a different coefficient.
    bivector2 = {('e2', 'e3'): 2.0}  # A bivector with a non-unit
    ↪    coefficient.
    theta2 = math.pi / 3  # 60 degrees rotation.

    rotor2 = compute_rotor(bivector2, theta2)
    print("\nRotor for bivector {('e2','e3'): 2.0} with 60°
    ↪    rotation:")
    print(multivector_to_string(rotor2))

    # Example 3: Additional demonstration using a bivector composed
    ↪    of multiple blades.
    bivector3 = {('e1', 'e2'): 0.6, ('e2', 'e3'): 0.8}
    theta3 = math.pi / 4  # 45 degrees rotation.

    rotor3 = compute_rotor(bivector3, theta3)
    print("\nRotor for bivector with components {('e1','e2'): 0.6,
    ↪    ('e2','e3'): 0.8} and 45° rotation:")
```

```
print(multivector_to_string(rotor3))
```

Chapter 38

Commutators in GA: Algebraic Structures Explored

Mathematical Definition and Algebraic Properties

The commutator in geometric algebra serves as a measure of the non-commutativity of the geometric product. For two multivectors A and B, the commutator is defined as

$$[A, B] = AB - BA,$$

where the product represents the geometric product. Unlike scalar multiplication, the product of multivectors generally does not commute. This non-commutativity is encoded in the antisymmetric nature of the commutator and plays a central role in the algebraic formulation of rotations and symmetry transformations. In many instances, the commutator defines a Lie bracket that satisfies the Jacobi identity, thereby linking the structure of geometric algebra with the Lie algebras that underpin various physical theories.

An extended variant of the commutator takes into account the grades of the multivectors. In such cases, the graded commutator is expressed as

$$[A, B]_\mathrm{G} = AB - (-1)^{\mathrm{grade}(A)\,\mathrm{grade}(B)} BA,$$

which refines the behavior of the bracket under transformations that mix even and odd grade elements. Nonetheless, the primary definition
$$[A, B] = AB - BA$$
remains central to many computational and theoretical applications.

Physical Relevance in Symmetry and Dynamics

Within physical theories, commutators are essential in characterizing the infinitesimal generators of continuous transformations. In the realm of quantum mechanics, for instance, the commutator of operators is directly related to uncertainty relations and the evolution of quantum states. Within the framework of geometric algebra, the commutator of bivectors emerges as an operator that generates rotations, mapping directly to the classical Lie algebra associated with the rotation group.

The algebraic properties encoded in the commutator provide a rich language for expressing conservation laws and symmetry transformations. In particular, the commutator structure encapsulates the behavior of angular momentum and other conserved quantities. By treating the geometric product as a fundamental operation and deriving the commutator from it, the framework coalesces algebraic precision with geometric intuition—thus serving as a powerful tool for the theoretical modeling and numerical simulation of dynamical systems.

Computational Implementation of Commutators

Accurate evaluation of commutators in a computational setting requires explicit calculation of the geometric products involved, followed by their subtraction. Multivectors are frequently represented as dictionaries mapping basis blade identifiers to their scalar coefficients. In this representation, the computation of the commutator demands careful aggregation of terms since the geometric product may yield contributions across a range of grades.

A representative implementation involves a function that accepts two multivectors along with a helper function for computing their geometric product. The following code snippet details a single Python function that computes the commutator of two multivectors. The function is designed to integrate seamlessly into larger computational frameworks where efficient evaluation of the geometric product is a critical prerequisite.

```python
def compute_commutator(A, B, geometric_product):
    """
    Compute the commutator of two multivectors A and B in the
    ↪ framework of geometric algebra.

    The commutator is defined as:
        [A, B] = A*B - B*A,
    where * denotes the geometric product.

    Args:
        A (dict): Multivector represented as a dictionary mapping
        ↪ basis blades (tuples) to coefficients.
        B (dict): Multivector represented similarly to A.
        geometric_product (function): A function that computes the
        ↪ geometric product of two multivectors.

    Returns:
        dict: The multivector resulting from the commutator [A, B].
    """
    # Compute the geometric product A*B.
    AB = geometric_product(A, B)

    # Compute the geometric product B*A.
    BA = geometric_product(B, A)

    # Form the commutator by subtracting BA from AB.
    commutator = {}
    for blade in set(AB.keys()).union(BA.keys()):
        commutator[blade] = AB.get(blade, 0) - BA.get(blade, 0)

    return commutator
```

This computational routine first evaluates the geometric products AB and BA using the provided function and then computes their difference. The function efficiently handles the multivector representations by iterating over the union of the basis blades occurring in either product. Such modular design promotes numerical efficiency and preserves the algebraic structure, which is indispensable when simulating systems where commutators play a pivotal role.

Python Code Snippet

```python
def count_inversions(lst):
    """
    Counts the number of inversions in a list.
    An inversion is a pair (i, j) such that i < j and lst[i] >
     lst[j].
    This is used to determine the sign change when reordering basis
     blades.
    """
    count = 0
    n = len(lst)
    for i in range(n):
        for j in range(i + 1, n):
            if lst[i] > lst[j]:
                count += 1
    return count

def multiply_basis_blades(b1, b2):
    """
    Multiply two basis blades represented as tuples of indices.

    The multiplication is performed by:
        1. Concatenating the indices from b1 and b2.
        2. Counting inversions in the combined list to determine the
         overall sign.
        3. Sorting the indices to achieve canonical order.
        4. Removing duplicate indices (each duplicate pair,
           corresponding to a squared basis vector in a Euclidean
         metric, cancels out).

    Args:
        b1 (tuple): First basis blade (e.g., (1,) represents e1).
        b2 (tuple): Second basis blade.

    Returns:
        tuple: A 2-tuple (result_blade, sign) where result_blade is
         a tuple
            representing the canonical blade, and sign is +1 or
             -1.
    """
    # Step 1: Concatenate the basis indices from b1 and b2.
    combined = list(b1) + list(b2)

    # Step 2: Count inversions to determine the sign factor.
    inversions = count_inversions(combined)
    sign = (-1) ** inversions

    # Step 3: Sort to obtain canonical order.
    sorted_combined = sorted(combined)
```

```python
    # Step 4: Remove duplicate indices (since e_i*e_i = 1 in
    ↪ Euclidean spaces).
    result = []
    i = 0
    while i < len(sorted_combined):
        if i < len(sorted_combined) - 1 and sorted_combined[i] ==
        ↪ sorted_combined[i + 1]:
            # Duplicate found; skip both indices.
            i += 2
        else:
            result.append(sorted_combined[i])
            i += 1
    return tuple(result), sign

def geometric_product(A, B):
    """
    Compute the geometric product of two multivectors A and B.

    Multivectors are represented as dictionaries mapping basis
    ↪ blades (tuples)
    to their scalar coefficients. For example, a scalar 3 can be
    ↪ written as {(): 3},
    and a vector e1 as {(1,): 1}.

    The function computes the product by multiplying every term in A
    ↪ with every term in B,
    using the multiplication rules for basis blades.

    Args:
        A (dict): The first multivector.
        B (dict): The second multivector.

    Returns:
        dict: The resulting multivector from the geometric product.
    """
    result = {}
    for blade_a, coeff_a in A.items():
        for blade_b, coeff_b in B.items():
            new_blade, factor = multiply_basis_blades(blade_a,
            ↪ blade_b)
            new_coeff = coeff_a * coeff_b * factor
            result[new_blade] = result.get(new_blade, 0) + new_coeff
    return result

def compute_commutator(A, B, geometric_product):
    """
    Compute the standard commutator [A, B] = A*B - B*A of two
    ↪ multivectors.

    Args:
        A (dict): The first multivector.
        B (dict): The second multivector.
```

            geometric_product (function): A function that computes the
            ↪   geometric product.

        Returns:
            dict: The resulting multivector representing the commutator
            ↪   [A, B].
        """
        # Compute A*B and B*A.
        AB = geometric_product(A, B)
        BA = geometric_product(B, A)

        # Subtract BA from AB, aggregating coefficients for each basis
        ↪   blade.
        commutator = {}
        for blade in set(AB.keys()).union(BA.keys()):
            commutator[blade] = AB.get(blade, 0) - BA.get(blade, 0)
        return commutator

def get_grade(multivector):
    """
    Retrieve the grade of a homogeneous multivector.
    Assumes that the multivector is homogeneous, i.e., all nonzero
    ↪   terms have the same grade.

    Args:
        multivector (dict): A multivector represented as a
        ↪   dictionary.

    Returns:
        int: The grade (number of basis vectors) of the first
        ↪   nonzero term.
            Scalars have grade 0.
    """
    for blade, coeff in multivector.items():
        if coeff != 0:
            return len(blade)
    return 0

def graded_commutator(A, B, geometric_product):
    """
    Compute the graded commutator of two homogeneous multivectors A
    ↪   and B.

    The graded commutator is defined as:
        [A, B]_G = A*B - (-1)^(grade(A)*grade(B)) B*A,
    which takes into account the signs due to the grades of A and B.

    Args:
        A (dict): The first homogeneous multivector.
        B (dict): The second homogeneous multivector.
        geometric_product (function): Function to compute the
        ↪   geometric product.

272

```
    Returns:
        dict: The resulting multivector from the graded commutator.
    """
    grade_A = get_grade(A)
    grade_B = get_grade(B)
    factor = (-1) ** (grade_A * grade_B)
    AB = geometric_product(A, B)
    BA = geometric_product(B, A)
    result = {}
    for blade in set(list(AB.keys()) + list(BA.keys())):
        result[blade] = AB.get(blade, 0) - factor * BA.get(blade, 0)
    return result

if __name__ == '__main__':
    # Define example multivectors.
    # Let A = 2 + e1, represented as {(): 2, (1,): 1}
    A = {(): 2, (1,): 1}
    # Let B = 3*e1 + 4*e2, represented as {(1,): 3, (2,): 4}
    B = {(1,): 3, (2,): 4}

    # Compute the geometric products A*B and B*A.
    product_AB = geometric_product(A, B)
    product_BA = geometric_product(B, A)

    # Compute the standard commutator [A, B] = A*B - B*A.
    comm = compute_commutator(A, B, geometric_product)

    # For homogeneous multivectors, demonstrate the graded
    ↪ commutator.
    # Let e1 = {(1,): 1} and e2 = {(2,): 1} be unit vectors.
    A_vec = {(1,): 1}
    B_vec = {(2,): 1}
    grad_comm = graded_commutator(A_vec, B_vec, geometric_product)

    # Display the results.
    print("Multivector A =", A)
    print("Multivector B =", B)
    print("\nGeometric Product A*B =", product_AB)
    print("Geometric Product B*A =", product_BA)
    print("\nCommutator [A, B] =", comm)
    print("\nGraded Commutator [e1, e2]_G =", grad_comm)
```

Chapter 39

Anti-Commutators: Properties and Physical Interpretations

Mathematical Formulation and Algebraic Characteristics

In geometric algebra, the anti-commutator is defined for any two multivectors A and B as

$$\{A, B\} = AB + BA,$$

where AB denotes the geometric product. This symmetric combination extracts the even (symmetric) component of the product, and its intrinsic symmetry is manifest in the equality

$$\{A, B\} = \{B, A\}.$$

For homogeneous elements, it is often the case that the anti-commutator isolates the scalar and higher-grade symmetric contributions inherent in the geometric product. For example, when restricting the discussion to vectors, the geometric product decomposes as

$$ab = a \cdot b + a \wedge b,$$

and the anti-commutator yields

$$\{a, b\} = ab + ba = 2\,a \cdot b,$$

thereby recovering twice the inner product. This characteristic underpins many algebraic manipulations and serves to reinforce the prominent role of symmetric operations in both computational routines and theoretical analyses.

Implications in Symmetry and Transformation Analysis

The symmetric nature of the anti-commutator has significant implications in the study of symmetry operations and invariance principles. In systems where transformations are represented through geometric products, the anti-commutator contributes exclusively to determining the symmetric, or Jordan, structure of the algebra. Such structures are crucial when reflecting on the invariance of bilinear forms, where the symmetric combination preserves parity under basis exchange.

In the context of physical theories, operators often require a symmetric combination to model observables that must remain invariant under specific transformations. The anti-commutator thus plays a central role in discussions regarding conserved quantities, where it frequently appears in the theoretical formulations that ensure the proper handling of symmetry-related constraints. The analysis of these symmetric components aids in the classification of multivector elements according to their transformation properties, which is essential for the rigorous formulation of physical laws.

Computational Implementation of Anti-Commutators in Geometric Algebra

When addressing the computational evaluation of anti-commutators, multivectors are typically represented as dictionaries mapping basis blades (expressed as ordered tuples) to their corresponding coefficients. The strategy involves independently computing the geometric products AB and BA, and then forming their sum to yield the anti-commutator. This procedure can be encapsulated in a single function.

The following Python function computes the anti-commutator $\{A, B\} = AB + BA$ for two multivectors. It leverages a helper function, `geometric_product`, which is assumed to perform the multiplication according to the algebraic rules of geometric algebra. The function aggregates the contributions from each product by iterating over the union of the basis blades present in both terms.

```python
def compute_anticommutator(A, B, geometric_product):
    """
    Compute the anti-commutator {A, B} = A*B + B*A for two
      multivectors A and B.

    Args:
        A (dict): A multivector represented as a dictionary mapping
          basis blades
                  (tuples) to their scalar coefficients.
        B (dict): A multivector represented in the same manner as A.
        geometric_product (function): A function that computes the
          geometric product
                                       of two multivectors.

    Returns:
        dict: A multivector representing the anti-commutator {A, B}.
    """
    AB = geometric_product(A, B)
    BA = geometric_product(B, A)
    result = {}
    for blade in set(AB.keys()).union(BA.keys()):
        result[blade] = AB.get(blade, 0) + BA.get(blade, 0)
    return result
```

This implementation underscores the linearity and symmetry of the anti-commutator operation. By cleanly separating the geometric products into their constituent terms, the function preserves the algebraic structure inherent in the multivector representation. Such computational strategies are essential for accurate numerical simulations and for verifying the symmetry properties that underlie many theoretical formulations in physics.

Python Code Snippet

```python
import itertools

def sign_of_permutation(lst):
    """
    Compute the sign (+1 or -1) of the permutation required to sort
      the list.
```

```
    This simple bubble sort based algorithm counts the number of
    ↪ swaps and
    returns +1 if even, -1 if odd.
    """
    arr = list(lst)
    swaps = 0
    n = len(arr)
    for i in range(n):
        for j in range(0, n - i - 1):
            if arr[j] > arr[j + 1]:
                arr[j], arr[j + 1] = arr[j + 1], arr[j]
                swaps += 1
    return 1 if swaps % 2 == 0 else -1

def remove_duplicates(sorted_blade):
    """
    Remove duplicates in pairs from a sorted list representing basis
    ↪ vector indices.
    In a Euclidean metric, the square of each basis vector is 1 so
    ↪ that e_i * e_i = 1.
    This function eliminates pairs (duplicates) and returns the
    ↪ resulting basis blade
    as a tuple.
    """
    result = []
    i = 0
    n = len(sorted_blade)
    while i < n:
        if i < n - 1 and sorted_blade[i] == sorted_blade[i + 1]:
            # Skip the pair since e_i * e_i = 1.
            i += 2
        else:
            result.append(sorted_blade[i])
            i += 1
    return tuple(result)

def multiply_blades(blade1, blade2):
    """
    Multiply two basis blades represented as tuples.
    The geometric product for basis blades in an orthonormal
    ↪ Euclidean space is given by:
        e_A * e_B = sign * e_C,
    where the result blade e_C is formed by the sorted concatenation
    ↪ of the indices in blade1 and blade2
    with duplicate entries removed in pairs, and sign is determined
    ↪ by the permutation required
    to sort the combined list.

    Args:
        blade1 (tuple): A tuple representing the basis indices of
        ↪ the first blade.
        blade2 (tuple): A tuple representing the basis indices of
        ↪ the second blade.
```

```
Returns:
    (int, tuple): A tuple (sign, result_blade) where sign is +1
    ↪ or -1, and result_blade is the
        resulting basis blade represented as an
        ↪ ordered tuple.
"""
combined = list(blade1) + list(blade2)
# Calculate the sign associated with reordering
permutation_sign = sign_of_permutation(combined)
sorted_combined = sorted(combined)
# Remove duplicate basis vectors (pairs) as e_i*e_i = 1.
result_blade = remove_duplicates(sorted_combined)
return permutation_sign, result_blade

def geometric_product(A, B):
    """
    Compute the geometric product of two multivectors A and B.
    Multivectors are represented as dictionaries mapping basis
    ↪ blades (tuples) to coefficients.

    The geometric product is performed term-by-term using the
    ↪ multiplication of individual
    basis blades based on the algorithm defined in multiply_blades.

    Args:
        A (dict): First multivector, e.g., { (1,): 3.0, (2,): -2.0 }
        B (dict): Second multivector.

    Returns:
        dict: A multivector (dictionary) representing the product.
    """
    result = {}
    for blade1, coeff1 in A.items():
        for blade2, coeff2 in B.items():
            sign, res_blade = multiply_blades(blade1, blade2)
            product_coeff = coeff1 * coeff2 * sign
            # Sum contributions to the same basis blade.
            if res_blade in result:
                result[res_blade] += product_coeff
            else:
                result[res_blade] = product_coeff
    # Eliminate terms with negligible coefficients.
    result = {blade: coeff for blade, coeff in result.items() if
    ↪ abs(coeff) > 1e-12}
    return result

def compute_anticommutator(A, B):
    """
    Compute the anti-commutator {A, B} defined as:
        {A, B} = A*B + B*A
    for two multivectors A and B.
```

```
    This function uses the geometric_product function to compute the
    ↪  two products
    and then sums them. The anti-commutator isolates the symmetric
    ↪  part of the
    geometric product and, for example, recovers 2 * (inner product)
    ↪  for vector inputs.

    Args:
        A (dict): Multivector A represented as a dictionary.
        B (dict): Multivector B represented in the same fashion.

    Returns:
        dict: A multivector representing the anti-commutator {A, B}.
    """
    AB = geometric_product(A, B)
    BA = geometric_product(B, A)
    result = {}
    for blade in set(AB.keys()).union(BA.keys()):
        result[blade] = AB.get(blade, 0) + BA.get(blade, 0)
    # Remove terms with small magnitude.
    result = {blade: coeff for blade, coeff in result.items() if
    ↪  abs(coeff) > 1e-12}
    return result

# Example usage:
if __name__ == "__main__":
    # Define two vectors a and b in a 3D Euclidean space.
    # Vectors are represented as multivectors with single basis
    ↪  elements:
    # a = 3 e1 - 2 e2 + 1.5 e3
    # b = 1 e1 + 4 e2 - 1 e3
    a = { (1,): 3.0, (2,): -2.0, (3,): 1.5 }
    b = { (1,): 1.0, (2,): 4.0, (3,): -1.0 }

    # Compute the geometric product A*B.
    ab = geometric_product(a, b)
    print("Geometric Product A*B:")
    for blade, coeff in sorted(ab.items()):
        print(f"{blade}: {coeff}")

    # Compute the anti-commutator {A, B} = A*B + B*A.
    anticomm = compute_anticommutator(a, b)
    print("\nAnti-Commutator {A, B} = A*B + B*A:")
    for blade, coeff in sorted(anticomm.items()):
        print(f"{blade}: {coeff}")

    # For vectors, the geometric product decomposes into a scalar
    ↪  (inner product)
    # and a bivector (outer product). The anti-commutator yields
    ↪  twice the inner product.
    dot_product = a.get((1,), 0) * b.get((1,), 0) + \
                  a.get((2,), 0) * b.get((2,), 0) + \
                  a.get((3,), 0) * b.get((3,), 0)
```

```
print("\n2 * Dot Product (from anti-commutator for vectors):")
print(2 * dot_product)
```

Chapter 40

Differential Operators in GA: A Computational Framework

Mathematical Foundations of Differential Calculus in Geometric Algebra

Differential operators in geometric algebra are constructed from the synthesis of standard calculus with the graded structure of multivectors. In this framework, the derivative is not restricted to scalar or vector functions but extends naturally to multivector fields. A fundamental operator in this context is the vector derivative, typically defined as

$$\partial = \sum_{i=1}^{n} e_i \frac{\partial}{\partial x_i},$$

where each e_i represents a grade-1 basis element and $\frac{\partial}{\partial x_i}$ denotes the partial derivative with respect to the coordinate x_i. This definition encapsulates the familiar operators of gradient, divergence, and curl into a single unified derivative, which, when applied via the geometric product, produces decompositions into symmetric and antisymmetric parts. The elegance of this approach lies in its capacity to act on a general multivector function while preserving the intrinsic algebraic structure.

The Vector Derivative in Geometric Algebra

Within geometric algebra, the vector derivative ∂ plays a central role in the differentiation of multivector functions. Its application is governed by well-established rules analogous to the Leibniz and chain rules of classical calculus. When ∂ multiplies a multivector function $F(x)$, the product is defined by the geometric product, ensuring that both inner and outer derivatives are present in the result. In particular, the derivative of a product FG is given by

$$\partial(FG) = (\partial F)G + F(\partial G),$$

thus maintaining the distributive and associative properties of the underlying algebra. The combination of these operations leads to representations that naturally generalize familiar calculus operations to settings where the objects possess a rich geometric interpretation.

Algorithmic Strategies for Computing Derivatives of Multivector Fields

The computational evaluation of derivatives in geometric algebra necessitates the systematic handling of basis blades and their associated coefficient functions. The approach involves decomposing a multivector field into its constituent parts, differentiating each coefficient function with respect to the relevant variables, and then reassembling the differentiated components into the original algebraic structure. Efficient algorithms must account for the linearity of differentiation and the inherent multiplicative relationships among basis elements.

A representative computational strategy involves a function that accepts a multivector field and a specified variable, and then returns the differentiated field. This function must iterate over each term in the field, apply the appropriate differentiation rules to the coefficients, and reconstruct the result in accordance with the algebraic properties of the basis blades. The following code snippet illustrates a prototype for such a function:

```
def differentiate_multivector(mv_field, var):
    """
    Compute the derivative of a multivector field with respect to a
    ↪ specified variable.

    Args:
        mv_field (dict): A multivector field represented as a
        ↪ dictionary that maps tuples
                        (basis blades) to coefficient functions
                        ↪ expressed in terms of
                        symbolic or numerical variables.
        var (str): The variable with respect to which
        ↪ differentiation is to be performed.

    Returns:
        dict: A new multivector field where each coefficient is
        ↪ replaced by its derivative
            with respect to 'var', preserving the structure of
            ↪ basis blades.
    """
    # Implementation will be provided later.
    pass
```

This function prototype embodies the essential steps required for the computation: identification of the multivector components, application of differentiation rules to each coefficient, and aggregation of the resulting terms while preserving the blade structure.

Computational Considerations and Efficiency

The practical implementation of differential operators in geometric algebra requires careful attention to data structures and algorithmic efficiency. Multivector fields are most naturally represented by sparse dictionary structures, where keys correspond to ordered tuples of basis indices and values represent the associated coefficient functions. Such a representation minimizes memory usage and allows rapid lookup and manipulation of terms.

The differentiation process must ensure that all algebraic properties are preserved. In particular, the product rule must be applied in a manner that respects the non-commutative nature of the geometric product. The numerical stability of the resulting computations can be enhanced by eliminating negligible contributions to the coefficients, and symbolic manipulation algorithms can leverage computer algebra systems to simplify the resultant expressions. Optimization of these routines is often crucial when these differential operators are to be embedded within larger simulation frameworks or iterative numerical solvers.

Efficient computation also involves parallelizing the differentiation over the independent components of the multivector field when applicable, as well as integrating caching strategies for repeated evaluations of common derivative structures. The interplay between symbolic and numerical techniques is central to the flexible and robust computation of differential operators in the context of geometric algebra.

Python Code Snippet

```
from sympy import symbols, diff, simplify
from collections import defaultdict, Counter

def multiply_blades(blade1, blade2):
    """
    Multiply two basis blades in an orthonormal Euclidean space.
    Each blade is represented as a tuple of integers (basis
    ↪ indices).
    The geometric product is computed by concatenating the two
    ↪ blades,
    counting the sign from reordering, and then canceling duplicate
    ↪ indices
    (since e_i*e_i = 1).
    """
    # Concatenate the two blades.
    combined = list(blade1) + list(blade2)

    # Compute the sign by counting inversions required to sort the
    ↪ list.
    sign = 1
    temp = combined[:]
    n = len(temp)
    for i in range(n):
        for j in range(i+1, n):
            if temp[i] > temp[j]:
                sign = -sign

    # Sort the combined list to bring it into a canonical order.
    sorted_combined = sorted(temp)

    # Remove pairs of repeated indices (each squared basis gives 1).
    # A basis index that appears an even number of times will
    ↪ cancel.
    blade_counter = Counter(sorted_combined)
```

```python
        result = []
        for k in sorted(blade_counter.keys()):
            if blade_counter[k] % 2 == 1:
                result.append(k)
    return tuple(result), sign

def geometric_product(mv1, mv2):
    """
    Compute the geometric product of two multivectors.

    Parameters:
      mv1, mv2: dictionaries mapping basis blade (tuple) to a sympy
      ↪ expression (coefficient).

    The function iterates over all terms in mv1 and mv2, multiplies
    ↪ the
    coefficients and the corresponding basis blades (using
    ↪ multiply_blades),
    and accumulates the result in a new multivector.
    """
    result = defaultdict(lambda: 0)
    for blade1, coeff1 in mv1.items():
        for blade2, coeff2 in mv2.items():
            new_blade, sign = multiply_blades(blade1, blade2)
            prod_coeff = simplify(coeff1 * coeff2 * sign)
            result[new_blade] += prod_coeff
    # Remove zero coefficient terms.
    result = {blade: coeff for blade, coeff in result.items() if
    ↪ coeff != 0}
    return result

def differentiate_multivector(mv_field, var):
    """
    Differentiate a multivector field with respect to a given
    ↪ variable.

    Parameters:
      mv_field: dictionary mapping basis blades (tuples) to sympy
      ↪ expressions.
      var: sympy symbol representing the variable to differentiate
      ↪ with respect to.

    Each coefficient function in the multivector is differentiated.
    """
    new_mv_field = {}
    for blade, coeff in mv_field.items():
        new_coeff = simplify(diff(coeff, var))
        if new_coeff != 0:
            new_mv_field[blade] = new_coeff
    return new_mv_field

def add_multivectors(mv1, mv2):
    """
```

```
    Add two multivectors represented as dictionaries.
    """
    result = mv1.copy()
    for blade, coeff in mv2.items():
        if blade in result:
            result[blade] = simplify(result[blade] + coeff)
        else:
            result[blade] = coeff
    # Remove any terms with zero coefficient.
    result = {blade: coeff for blade, coeff in result.items() if
↪       coeff != 0}
    return result

def scale_multivector(mv, scalar):
    """
    Multiply a multivector by a scalar.
    """
    return {blade: simplify(coeff * scalar) for blade, coeff in
↪       mv.items() if simplify(coeff * scalar) != 0}

# Example usage and demonstration of key GA differential operations.
if __name__ == "__main__":
    # Define symbolic variables.
    x, y = symbols('x y')

    # Define a multivector F as a dictionary.
    # Example: F = 1 + x*e1 + y*e2 + 1*(e1^e2)
    # Here, scalar part is represented by the empty tuple ()
    # e1 is represented by (1,), e2 by (2,), and e1^e2 by (1,2)
↪   (with the convention that the tuple is ordered).
    F = {
        (): 1,          # scalar 1
        (1,): x,        # x * e1
        (2,): y,        # y * e2
        (1,2): 1        # 1 * (e1 ^ e2)
    }

    # Differentiate F with respect to x.
    dF_dx = differentiate_multivector(F, x)
    print("dF/dx =", dF_dx)

    # Define another multivector G.
    # Example: G = (x+y) + 2*e1
    G = {
        (): x + y,      # scalar part: (x + y)
        (1,): 2         # 2 * e1
    }

    # Compute the geometric product H = F * G.
    H = geometric_product(F, G)
    print("Geometric product H = F * G:", H)

    # Demonstrate the product rule for differentiation:
```

```
# d/dx (F * G) = (dF/dx) * G + F * (dG/dx)
dG_dx = differentiate_multivector(G, x)
left_side = differentiate_multivector(H, x)
right_side =
 ↪  add_multivectors(geometric_product(differentiate_multivector(F,
 ↪  x), G),
                                    geometric_product(F, dG_dx))

print("d/dx (F * G) =", left_side)
print("d/dx (F)*G + F*d/dx (G) =", right_side)
```

Chapter 41

Multivector Derivatives: Algorithms and Applications

Mathematical Formulation of Multivector Derivatives

Multivector functions are defined on a manifold and take values in a graded algebra with basis elements of varying grades. In this formalism, the derivative is extended beyond scalar or vector fields to encompass the full algebraic structure. Given a multivector function $F(x)$ defined on a coordinate domain, the derivative is implemented via a vector derivative operator defined as

$$\partial = \sum_{i=1}^{n} e_i \frac{\partial}{\partial x_i},$$

where every e_i denotes a grade-1 basis element and $\frac{\partial}{\partial x_i}$ represents the partial derivative with respect to the coordinate x_i. The application of ∂ to $F(x)$ inherently produces a combination of inner and outer parts. This decomposition is a consequence of the geometric product that underlies the multivector structure and faithfully reproduces common differential operators, thereby unifying the gradient, divergence, and curl within a single framework.

The algebraic behavior under differentiation is characterized by a generalized product rule. For two multivector functions $F(x)$ and $G(x)$, differentiation is conducted as

$$\partial(F\,G) = (\partial F)\,G + F\,(\partial G).$$

This formula maintains the distributive nature inherent in the geometric product while accounting for the graded structure of the algebra. The derived expressions preserve the geometric interpretation of the fields even after differentiation, ensuring that both scalar and higher-grade components are treated in a consistent manner.

Algorithmic Implementation of Multivector Differentiation

The computational differentiation of a multivector field is an algorithmic process that leverages the linearity of differentiation and the discrete representation of the algebra. In many computational settings, a multivector field is encoded as a dictionary where each key is a tuple corresponding to a specific basis blade and its associated value is a scalar coefficient function. Differentiation is performed by iterating through the structure, differentiating each coefficient with respect to a specified variable, and reconstructing the multivector to preserve its algebraic structure.

An exemplary function illustrates this approach. The function processes the multivector field by applying symbolic differentiation to every coefficient. The resulting field maintains a sparse representation through the removal of null coefficients. The code snippet below demonstrates an implementation of this technique:

```
def differentiate_multivector(mv_field, var):
    """
    Compute the derivative of a multivector field with respect to a
    ↪ given variable.

    The multivector field is represented as a dictionary mapping
    ↪ basis blades (tuples)
    to coefficient functions defined in a symbolic framework. This
    ↪ function iterates
    through each basis blade, differentiates the corresponding
    ↪ coefficient with respect
    to the variable 'var', and reconstructs the multivector field,
    ↪ preserving the underlying
```

```
algebraic structure by discarding any terms with a zero
↪ coefficient.
"""
new_mv_field = {}
for blade, coeff in mv_field.items():
    new_coeff = simplify(diff(coeff, var))
    if new_coeff != 0:
        new_mv_field[blade] = new_coeff
return new_mv_field
```

This implementation emphasizes the abstraction of the differentiation process away from the complexity of rearranging basis elements. The algorithm maintains modularity through the independent processing of each multivector component, thereby permitting efficient scaling for high-dimensional algebras. By focusing solely on the coefficient functions, the core properties of differentiation are preserved regardless of the underlying blade structure.

Applications in Computational Modeling

Multivector derivatives play a pivotal role in computational modeling of physical systems, where problems are articulated in terms of field equations possessing intricate geometric content. In electromagnetism, fluid dynamics, and quantum mechanics, spatial and temporal variations of multivector fields are studied using differential operators that account for both scalar and bivector (or higher-grade) contributions. The differentiation algorithms provide a robust tool for evaluating such variations with precision.

Through the combination of symbolic and numerical techniques, the derivatives serve as the foundation for constructing discrete event solvers and iterative refinement routines. The algorithms allow for the seamless integration of multivector calculus into simulation frameworks, enabling automated differentiation that rigorously adheres to the underlying structure of the algebra. This approach supports a broad spectrum of applications, ranging from the computation of conserved quantities in field theories to the evaluation of gradients in optimization problems embedded in physical simulations.

The development of these techniques ensures that multivector differentiation is both computationally efficient and mathematically rigorous. Their adoption in computational physics and computer science has demonstrated the capacity to simplify the handling of complex derivative expressions while preserving core geometric intuitions.

Python Code Snippet

```
from sympy import symbols, diff, simplify, sin, cos, exp

def differentiate_multivector(mv_field, var):
    """
    Compute the derivative of a multivector field with respect to a
    ↪ given variable.

    The multivector field is represented as a dictionary mapping
    ↪ basis blades (tuples)
    to coefficient functions (sympy expressions). This function
    ↪ iterates through each blade,
    differentiates its coefficient with respect to 'var', simplifies
    ↪ the result, and discards
    any terms that evaluate to zero.
    """
    new_mv_field = {}
    for blade, coeff in mv_field.items():
        new_coeff = simplify(diff(coeff, var))
        if new_coeff != 0:
            new_mv_field[blade] = new_coeff
    return new_mv_field

def geometric_product_blades(b1, b2):
    """
    Compute the geometric product of two basis blades represented as
    ↪ tuples.

    The geometric product for basis elements is defined via the
    ↪ anti-commutative property
    for distinct basis vectors (i.e. e_i e_j = - e_j e_i for i != j)
    ↪ and by the property
    e_i * e_i = 1 for an orthonormal metric. This function merges
    ↪ the two blades,
    calculates the sign from the reordering (number of swaps
    ↪ required to sort the combined list),
    and removes duplicate basis elements (since e_i*e_i = 1).

    For example, geometric_product_blades((1,), (1,2)) gives (1,2)
    ↪ with a positive sign.
    """
    # Combine the two blades
    blade_combined = list(b1) + list(b2)

    # Count the number of swaps required to sort the combined list.
    swaps = 0
```

```
        arr = list(blade_combined)
        for i in range(len(arr)):
            for j in range(i+1, len(arr)):
                if arr[i] > arr[j]:
                    swaps += 1
        sign = (-1)**swaps

        # Sort the blade and remove duplicate pairs (since e_i*e_i = 1).
        sorted_blade = sorted(blade_combined)
        result = []
        i = 0
        while i < len(sorted_blade):
            if i + 1 < len(sorted_blade) and sorted_blade[i] ==
            ↪   sorted_blade[i + 1]:
                # Remove the pair and omit from the resulting blade.
                i += 2
            else:
                result.append(sorted_blade[i])
                i += 1
        return sign, tuple(result)

    def geometric_product(mv1, mv2):
        """
        Compute the geometric product of two multivectors.

        Each multivector is a dictionary mapping basis blades (tuples)
        ↪   to sympy expressions.
        For every pair of blades from mv1 and mv2, the corresponding
        ↪   coefficients multiply,
        and the basis blades combine through the geometric product (with
        ↪   appropriate sign adjustments).
        The result is then simplified.
        """
        result = {}
        for blade1, coeff1 in mv1.items():
            for blade2, coeff2 in mv2.items():
                sign, blade_product = geometric_product_blades(blade1,
                ↪   blade2)
                term = sign * coeff1 * coeff2
                if blade_product in result:
                    result[blade_product] += term
                else:
                    result[blade_product] = term
        # Simplify coefficients in the resulting multivector.
        for blade in result:
            result[blade] = simplify(result[blade])
        return result

    def add_multivectors(mv1, mv2):
        """
        Add two multivectors represented as dictionaries. Where common
        ↪   basis blades occur,
        their coefficients are added.
```

```python
    """
    result = mv1.copy()
    for blade, coeff in mv2.items():
        if blade in result:
            result[blade] += coeff
        else:
            result[blade] = coeff
    # Simplify the coefficients.
    for blade in result:
        result[blade] = simplify(result[blade])
    return result

def print_multivector(mv):
    """
    Nicely print a multivector.

    The multivector is expected to be a dictionary mapping basis
    ↪ blades (tuples)
    to sympy expressions. The scalar part is printed as 1, and other
    ↪ blades are denoted
    by e1, e2, etc.
    """
    terms = []
    # Sort terms by grade and lexicographically.
    for blade, coeff in sorted(mv.items(), key=lambda item:
    ↪ (len(item[0]), item[0])):
        if blade == ():
            b_str = "1"   # Scalar part
        else:
            b_str = ''.join(['e{}'.format(i) for i in blade])
        terms.append(f"{coeff}*{b_str}")
    print(" + ".join(terms))

if __name__ == "__main__":
    # Define the symbolic variable.
    x = symbols('x')

    # Define an example multivector function F(x):
    # F(x) = x^2 (scalar) + sin(x)*e1 + cos(x)*e2 + exp(x)*e1e2
    F = {
        (): x**2,
        (1,): sin(x),
        (2,): cos(x),
        (1,2): exp(x)
    }

    # Define another multivector function G(x):
    # G(x) = cos(x) (scalar) + x*e1 - sin(x)*e2 + x^3*e1e2
    G = {
        (): cos(x),
        (1,): x,
        (2,): -sin(x),
        (1,2): x**3
```

```python
}

# Compute the derivative of F(x) with respect to x.
dF = differentiate_multivector(F, x)
print("Derivative of F(x):")
print_multivector(dF)

# Compute the derivative of G(x) with respect to x.
dG = differentiate_multivector(G, x)
print("\nDerivative of G(x):")
print_multivector(dG)

# Demonstrate the product rule:
# (d/dx)(F*G) = (dF)*G + F*(dG)
FG = geometric_product(F, G)
dFG_direct = differentiate_multivector(FG, x)
product_rule = add_multivectors(geometric_product(dF, G),
↪   geometric_product(F, dG))

print("\nDerivative of F*G computed directly:")
print_multivector(dFG_direct)

print("\nDerivative of F*G computed via product rule:")
print_multivector(product_rule)

# Validate the product rule by comparing both results.
diff_mv = {}
all_blades = set(list(dFG_direct.keys()) +
↪   list(product_rule.keys()))
for blade in all_blades:
    diff_coef = simplify(dFG_direct.get(blade, 0) -
↪       product_rule.get(blade, 0))
    if diff_coef != 0:
        diff_mv[blade] = diff_coef
print("\nDifference (should be zero if product rule holds):")
print_multivector(diff_mv)
```

Chapter 42

Differential Forms in GA: Integration and Differentiation

Mathematical Foundations of Differential Forms in GA

Differential forms, as antisymmetric multilinear mappings, are naturally embedded in geometric algebra through the identification of blades with oriented subspaces. In this framework, a differential form is expressed as a multivector

$$\Omega = \sum_{k=0}^{n} \omega_k,$$

where each ω_k is a grade-k component. The exterior, or wedge, product in conventional differential geometry is replaced by the outer product in geometric algebra. This identification permits the algebraic manipulation of differential forms without recourse to an explicit coordinate representation. Furthermore, the inherent antisymmetry of differential forms is preserved by the properties of the outer product, ensuring that for any two basis vectors e_i and e_j, the relation

$$e_i \wedge e_j = -e_j \wedge e_i$$

holds automatically.

Integration in the Framework of Geometric Algebra

The integration of differential forms within geometric algebra adopts the language of multivector integration. For a differential form Ω, integration over a manifold M is defined by a mapping

$$\int_M \Omega,$$

which, by virtue of the generalized Stokes theorem, relates the integration of the exterior derivative of a form to the integration of the original form over the boundary ∂M, namely,

$$\int_M d\Omega = \int_{\partial M} \Omega.$$

This approach unifies classical integration techniques with the higher-dimensional algebraic structures found in geometric algebra. The process of integration respects the graded structure of the algebra, allowing the simultaneous treatment of scalar, vector, and higher-grade components in a consistent manner. The use of differential forms proves advantageous in problems where orientation and multidimensional flux are of central importance.

Differentiation through the Exterior Derivative

The operation of differentiation in the context of differential forms is encapsulated by the exterior derivative operator d. In geometric algebra, this operator is characterized by its ability to raise the grade of a given form by one, while obeying the nilpotency property

$$d^2 = 0.$$

For a differential form Ω represented as a multivector, the exterior derivative is defined by

$$d\Omega = d \wedge \Omega,$$

where the derivative operator d is extended to act on each coefficient function present in Ω. This formulation generalizes the familiar gradient, divergence, and curl operations by embedding them within a unified algebraic structure. The exterior derivative adheres to a graded version of the Leibniz rule, ensuring that for differential forms Ω_1 and Ω_2,

$$d(\Omega_1 \wedge \Omega_2) = d\Omega_1 \wedge \Omega_2 + (-1)^{\text{grade}(\Omega_1)} \Omega_1 \wedge d\Omega_2.$$

This property is essential for maintaining consistency across varying grades, and it finds direct application in the formulation of conservation laws and the analysis of field theories.

Algorithmic Implementation: Computation of the Exterior Derivative

A computational approach to the exterior derivative necessitates a data structure capable of encoding differential forms via their constituent basis blades and associated coefficient functions. In practical implementations, a differential form is represented as a dictionary, in which keys are tuples corresponding to indices of basis vectors, and values are symbolic expressions representing the coefficient functions. The following function, implemented in Python with the aid of symbolic computation libraries, computes the exterior derivative of a differential form. The design of the function preserves antisymmetry by constructing new keys that reflect the requisite wedge product ordering.

```
def exterior_derivative(form, coords):
    """
    Compute the exterior derivative of a differential form within
        the framework of geometric algebra.

    The differential form is represented as a dictionary mapping
        tuples (representing the
    ordered indices of basis elements) to symbolic expressions. The
        'coords' parameter is a list
    of coordinate symbols with respect to which the partial
        derivatives are computed.

    The function iterates over each term in the form, applying the
        partial derivative with
    respect to each coordinate. The antisymmetry of the resulting
        derivative is ensured by
    inserting the coordinate index into the tuple and then sorting,
        which respects the wedge product
    structure. Terms that simplify to zero are omitted from the
        final result.
```

```
    Returns a new dictionary representation of the exterior
 ↪   derivative of the input form.
    """
    d_form = {}
    for idx, expr in form.items():
        for i, coord in enumerate(coords):
            # Create the new index tuple by merging and sorting to
            ↪   enforce antisymmetry.
            new_idx = tuple(sorted(idx + (i,)))
            d_form[new_idx] = d_form.get(new_idx, 0) +
            ↪   expr.diff(coord)
    # Simplify and remove zero terms.
    return {k: simplify(v) for k, v in d_form.items() if simplify(v)
    ↪   != 0}
```

Python Code Snippet

```
# ==================================================================
# Differential Forms in Geometric Algebra:
# Integration and Differentiation - Comprehensive Python
 ↪   Implementation
# ==================================================================

from sympy import symbols, diff, simplify, integrate, Function
import sympy as sp

def sort_with_sign(indices):
    """
    Sort the tuple 'indices' and compute the permutation signature.
    This function uses a simple bubble sort algorithm to count the
    ↪   number of transpositions
    required to reach the sorted order. It returns the sorted list
    ↪   of indices and the sign (+1 or -1).
    """
    indices = list(indices)
    sign = 1
    n = len(indices)
    # Bubble sort to determine permutation sign
    for i in range(n):
        for j in range(n - 1):
            if indices[j] > indices[j + 1]:
                indices[j], indices[j + 1] = indices[j + 1],
                ↪   indices[j]
                sign *= -1
    return indices, sign

def wedge_product(form1, form2):
    """
    Compute the wedge (exterior) product of two differential forms.
```

Each differential form is represented as a dictionary:
  - Keys: tuples corresponding to the ordered indices of the
  ↪ *basis elements.*
  - Values: the corresponding symbolic expressions.

The wedge product is performed term by term by merging the index
  ↪ *tuples. If any*
repeated indices occur in the merged tuple, the overall term
  ↪ *vanishes (enforcing antisymmetry).*
The function also computes the correct sign resulting from
  ↪ *sorting the merged indices.*
"""
```python
result = {}
for idx1, expr1 in form1.items():
    for idx2, expr2 in form2.items():
        # If there is any common index, the wedge product is
        ↪ zero.
        if set(idx1) & set(idx2):
            continue
        combined = idx1 + idx2
        sorted_indices, sign = sort_with_sign(combined)
        new_key = tuple(sorted_indices)
        new_expr = sign * expr1 * expr2
        if new_key in result:
            result[new_key] += new_expr
        else:
            result[new_key] = new_expr
# Simplify the result and remove any zero terms.
return {k: simplify(v) for k, v in result.items() if simplify(v)
    ↪ != 0}

def exterior_derivative(form, coords):
    """
```
Compute the exterior derivative of a differential form within
  ↪ *the Geometric Algebra framework.*

Parameters:
  form   : dict
            Differential form represented as a dictionary with
              ↪ *keys as tuples*
            of basis indices and values as symbolic expressions.
  coords : list
            List of coordinate symbols with respect to which
              ↪ *differentiation is carried out.*

The function implements the operator:
    $d = d$,
by differentiating each coefficient with respect to all
  ↪ *coordinates,*
inserting the corresponding coordinate index into the key, and
  ↪ *enforcing*
antisymmetry via sorting and a sign correction.

299

```
    """
    d_form = {}
    for idx, expr in form.items():
        for i, coord in enumerate(coords):
            # Append the derivative coordinate index to the existing
            ↪    key.
            new_tuple = idx + (i,)
            sorted_tuple, sign = sort_with_sign(new_tuple)
            new_key = tuple(sorted_tuple)
            new_expr = sign * diff(expr, coord)
            if new_key in d_form:
                d_form[new_key] += new_expr
            else:
                d_form[new_key] = new_expr
    # Simplify expressions and remove terms that vanish.
    return {k: simplify(v) for k, v in d_form.items() if simplify(v)
    ↪    != 0}

def integrate_top_form(form, coords, limits):
    """
    Integrate a top-degree differential form over a rectangular
    ↪    domain.

    Parameters:
      form   : dict
               Differential form represented as a dictionary. The
               ↪    top form is assumed to
               be identified by the sorted tuple (0, 1, ..., n-1),
               ↪    where n is the number of coordinates.
      coords : list
               List of coordinate symbols.
      limits : dict
               Dictionary mapping each coordinate symbol to a tuple
               ↪    (lower_limit, upper_limit).

    Returns:
      The value of the iterated integral of the top form component.

    Note: This function is primarily intended for forms in which the
    ↪    integration domain is
          a product of intervals (e.g., a cube or rectangular
          ↪    region).
    """
    top_key = tuple(range(len(coords)))
    if top_key not in form:
        raise ValueError("The provided form does not contain the
        ↪    top-degree component for integration.")
    expr = form[top_key]
    # Perform iterated integration over all coordinates using
    ↪    provided limits.
    for coord in coords:
        lower, upper = limits[coord]
        expr = integrate(expr, (coord, lower, upper))
```

```
        return simplify(expr)

def form_to_str(form):
    """
    Convert a differential form from dictionary representation to a
    ↪ human-readable string.

    This helper function constructs a string by enumerating each
    ↪ term and its corresponding
    basis element representation.
    """
    terms = []
    for indices in sorted(form.keys()):
        term = f"{form[indices]}"
        if indices:
            basis_str = "".join([f"dx_{i}" for i in indices])
            term += " " + basis_str
        terms.append(term)
    return " + ".join(terms)

# ================================================================
# Demonstration of the Differential Forms Algorithms
# ================================================================

if __name__ == "__main__":
    # Define coordinate symbols (e.g., for ³)
    x, y, z = symbols('x y z')
    coords = [x, y, z]

    # Define symbolic functions for demonstration
    f = Function('f')(x, y, z)
    g = Function('g')(x, y, z)
    h = Function('h')(x, y, z)

    # -------------------------------
    # Example 1: Exterior Derivative of a 0-form
    # -------------------------------
    # Represent a scalar function f(x,y,z) as a 0-form.
    zero_form = {(): f}    # 0-form is represented by the empty tuple.
    df = exterior_derivative(zero_form, coords)
    print("Exterior derivative of the 0-form f:")
    print(form_to_str(df))

    # -------------------------------
    # Example 2: Exterior Derivative of a 1-form
    # -------------------------------
    # Define the 1-form  = g dx + h dy.
    one_form = {(0,): g, (1,): h}
    domega = exterior_derivative(one_form, coords)
    print("\nExterior derivative of the 1-form  = g dx + h dy:")
    print(form_to_str(domega))

    # -------------------------------
```

```
# Example 3: Wedge Product of Differential Forms
# -------------------------------
# Compute the wedge product of the basis 1-forms dx and dy.
dx = {(0,): 1}
dy = {(1,): 1}
dx_wedge_dy = wedge_product(dx, dy)
print("\nWedge product dx  dy:")
print(form_to_str(dx_wedge_dy))

# -------------------------------
# Example 4: Integration of a Top-Degree Form
# -------------------------------
# Consider a 3-form on ³:  = f(x,y,z) dx  dy  dz.
three_form = {(0, 1, 2): f}
limits = {x: (0, 1), y: (0, 1), z: (0, 1)}
integral_result = integrate_top_form(three_form, coords, limits)
print("\nIntegration of the 3-form f dx  dy  dz over the unit
 ↪  cube [0,1]^3:")
print(integral_result)
```

Chapter 43

Lie Groups in GA: Connecting Algebra with Geometry

Foundations of Lie Groups and Geometric Algebra

Lie groups are smooth manifolds endowed with continuous group operations that are differentiable. In the context of geometric algebra (GA), these groups acquire representations as multivectors whose algebraic structure seamlessly encodes geometric transformations. For instance, continuous rotations in $\mathbb{R}^3$ are compactly represented by rotors, which are elements of GA constructed as the exponential of bivectors. The exponential mapping,

$$\exp(B) = \sum_{k=0}^{\infty} \frac{B^k}{k!},$$

where B is a bivector, establishes a direct correspondence between the Lie algebra and the Lie group. The bivector subspace of a geometric algebra serves as the natural arena for the generators of rotations, and the antisymmetric properties inherent in the outer product ensure that the Lie bracket defined by

$$[B_1, B_2] = B_1 B_2 - B_2 B_1$$

captures the non-commutative structure of the underlying symmetry group.

Representing Lie Groups in Geometric Algebra

In GA, a Lie group element can often be expressed in the form

$$R = \exp\left(-\frac{1}{2}\theta B\right),$$

where θ is the magnitude of the rotation and B is a unit bivector representing the plane of rotation. Such representations are not only elegant but also offer computational advantages. The group product is implemented via the geometric product, thereby rendering the composition of transformations both algebraically succinct and geometrically transparent. Moreover, the inverse of a group element is naturally acquired through the reversion operation in GA. The interplay between the exponential map and the geometric product facilitates an unambiguous translation of Lie group operations into the language of multivectors, unifying algebraic manipulations with geometric intuitions.

Code Implementation and Computational Aspects

The computational realization of Lie group representations in GA requires mapping symbolic Lie algebra elements onto their corresponding multivector forms. A typical task involves converting a Lie algebra element into a rotor representation by means of the exponential mapping. A function dedicated to this task encapsulates all necessary mathematical operations and serves as a building block for more comprehensive simulation routines. An illustrative single-function code snippet is presented below.

```
def lie_group_to_ga(lie_element):
    """
    Converts a Lie algebra element into its corresponding Geometric
    ↪ Algebra (GA) representation.

    The input lie_element is assumed to be a symbolic expression
    ↪ composed of bivector
```

```
            components corresponding to the Lie algebra generators. The
        ↪   function calculates
            the rotor using the exponential map:
                R = exp(-0.5 * theta * B)
            where theta represents the rotation angle and B the normalized
        ↪   bivector generator.

            Parameters:
                lie_element : symbolic expression
                    A Lie algebra element expressed in terms of its bivector
                ↪   basis.

            Returns:
                ga_element : multivector
                    The GA representation of the Lie group element as a
                ↪   rotor.
            """
            # Implementation details will be provided in the computational
        ↪   module.
            pass
```

The function encapsulates a vital conversion process, translating a Lie algebra element into a rotor that can be manipulated via the geometric product. The strategy ensures that all group operations, such as composition and inversion, are performed in a unified algebraic framework.

Algebraic Structure and Geometric Transformations

The geometric product in GA furnishes a direct mechanism for composing transformations. Consider two group elements represented as rotors,

$$R_1 = \exp\left(-\frac{1}{2}\theta_1 B_1\right) \quad \text{and} \quad R_2 = \exp\left(-\frac{1}{2}\theta_2 B_2\right),$$

whose product,

$$R = R_1 R_2,$$

encapsulates the successive application of rotations. The result R is itself a multivector whose structure preserves the compounded rotational effect. This inherent compatibility of algebra and geometry facilitates the use of GA for describing continuous symmetries and for simulating complex spatial transformations. In addition, the well-behaved nature of the exponential map in GA permits the retrieval of the infinitesimal generator from a composite transformation via the logarithmic map,

$$B = -2\log R,$$

thereby ensuring a coherent correspondence between the Lie group and its Lie algebra.

Analysis of Commutators and the Exponential Map in GA

The examination of Lie algebra commutators within GA is performed through the commutator operation provided by the geometric product. For bivectors B_1 and B_2, the commutator,

$$[B_1, B_2] = B_1 B_2 - B_2 B_1,$$

embodies the fundamental non-commutativity of rotational generators. This operation reflects the curvature of the manifold on which the Lie group acts. The exponentiation of bivectors to generate group elements, as indicated by the series expansion

$$\exp(B) = \sum_{k=0}^{\infty} \frac{B^k}{k!},$$

ensures that the transition from the algebraic to the geometric realm is smooth and analytically tractable. The precision in reproducing group properties, such as closure and invertibility, is maintained through careful manipulation of the bivector components, the reversion operation, and normalization processes intrinsic to GA. Such a synthesis of algebraic and geometric techniques exemplifies the potency of GA in modeling physical symmetry operations.

Python Code Snippet

```
import sympy as sp

# Define the symbolic variable for the rotation angle.
# Note: 'theta' will be used for series expansion purposes in some
↪ functions.
theta = sp.symbols('theta')
```

```python
#############################################
# Bivector Class Definition
#############################################
class Bivector:
    """
    A simple class to encapsulate a bivector element in geometric
    ↪ algebra.
    For demonstration, the bivector is represented as a sympy
    ↪ symbolic expression.
    """
    def __init__(self, expr):
        # expr represents the bivector symbol (or expression), e.g.,
        ↪ e12.
        self.expr = sp.simplify(expr)

    def __mul__(self, other):
        """
        Multiplication: Supports scalar multiplication or
        ↪ multiplication
        between bivectors. For two bivectors in an orthonormal
        ↪ basis, one
        typically has B^2 = -|B|^2 for normalized bivectors. Here we
        ↪ leave
        the product symbolic.
        """
        if isinstance(other, (int, float, sp.Expr)):
            return Bivector(self.expr * other)
        elif isinstance(other, Bivector):
            # The geometric product of bivectors is non-commutative;
            ↪ here we
            # simply return the symbolic product.
            return Bivector(sp.simplify(self.expr * other.expr))
        else:
            raise TypeError("Unsupported multiplication type for
            ↪ Bivector.")

    def __add__(self, other):
        if isinstance(other, Bivector):
            return Bivector(self.expr + other.expr)
        else:
            return Bivector(self.expr + other)

    def __pow__(self, n):
        """
        Raises the bivector to an integer power n.
        """
        if n == 0:
            return sp.Integer(1)
        result = self.expr
        for _ in range(1, n):
            result = sp.simplify(result * self.expr)
        return result
```

```
    def __str__(self):
        return f"Bivector({self.expr})"

#############################################
# Exponential of a Bivector using Series Expansion
#############################################
def exp_bivector(B, angle_val, n_terms=10):
    """
    Computes the exponential of a bivector using the series
    ↪ expansion.

    Given a bivector B, the rotor is defined as:
        R = exp(-1/2 * angle_val * B) = sum_{k=0}^{n_terms-1} [(-1/2
        ↪ * angle_val * B)^k] / k!

    Parameters:
        B         : An instance of Bivector.
        angle_val : The rotation angle (can be symbolic or numeric).
        n_terms   : Number of terms to include in the series
        ↪ expansion.

    Returns:
        A sympy expression approximating the rotor.
    """
    series_sum = 0
    for k in range(n_terms):
        term = ((-sp.Rational(1,2) * angle_val * B.expr)**k) /
        ↪ sp.factorial(k)
        series_sum += term
    return sp.simplify(series_sum)

#############################################
# Rotor Construction: Lie Algebra to Geometric Algebra Mapping
#############################################
def lie_group_to_ga(bivector_expr, angle_val, n_terms=10):
    """
    Converts a Lie algebra element (represented as a bivector) into
    ↪ its corresponding
    geometric algebra (GA) rotor using the exponential map.

    The rotor is given by:
        R = exp(-1/2 * angle_val * bivector_expr)

    Parameters:
        bivector_expr : A sympy expression representing the bivector
        ↪ (e.g., e12).
        angle_val     : The rotation angle.
        n_terms       : Series expansion terms.

    Returns:
        A sympy expression representing the GA rotor.
    """
```

```python
    B = Bivector(bivector_expr)
    R = exp_bivector(B, angle_val, n_terms)
    return sp.simplify(R)

##############################################
# Commutator of Two Bivectors
##############################################
def commutator(B1_expr, B2_expr):
    """
    Computes the commutator between two bivectors:

        [B1, B2] = B1 * B2 - B2 * B1

    Parameters:
        B1_expr, B2_expr : sympy expressions representing the
          ↪  bivector generators.

    Returns:
        A sympy expression of their commutator.
    """
    B1 = sp.sympify(B1_expr)
    B2 = sp.sympify(B2_expr)
    return sp.simplify(B1 * B2 - B2 * B1)

##############################################
# Rotor Reversion (Inverse) via the Exponential Map
##############################################
def rotor_reverse(bivector_expr, angle_val, n_terms=10):
    """
    Computes the reverse (or reversion) of a rotor.

    For a rotor defined by:
        R = exp(-1/2 * angle_val * bivector_expr),
    its reverse is given by:
        R_reverse = exp(+1/2 * angle_val * bivector_expr)

    This reverse corresponds to the inverse for unit rotors.

    Parameters:
        bivector_expr : A sympy expression representing the
          ↪  bivector.
        angle_val     : The rotation angle used in rotor
          ↪  construction.
        n_terms       : Series expansion terms.

    Returns:
        The reversed rotor as a sympy expression.
    """
    # Note: Since our rotor is constructed via the series expansion,
    # reversing is achieved by changing the sign of the rotation
      ↪  angle.
    return lie_group_to_ga(bivector_expr, -angle_val, n_terms)
```

```python
###########################################
# Rotor Composition: Geometric Product of Two Rotors
###########################################
def compose_rotors(R1, R2):
    """
    Composes two rotors by applying the geometric product.

    Parameters:
        R1, R2 : Sympy expressions representing GA rotors.

    Returns:
        A sympy expression corresponding to the composite rotor.
    """
    return sp.simplify(R1 * R2)


###########################################
# Logarithm of a Rotor (Retrieving the Lie Algebra Element)
###########################################
def rotor_log(R):
    """
    Approximates the logarithm of a rotor.

    For a rotor R = exp(-1/2 * angle * B), the logarithm ideally
    ↪  recovers:
        log(R) = -1/2 * angle * B.

    Here, we use sympy's logarithm for a symbolic approximation.

    Parameters:
        R : A sympy expression representing the rotor.

    Returns:
        The logarithm of R as a sympy expression.
    """
    return sp.simplify(sp.log(R))


###########################################
# Main Testing Routine
###########################################
if __name__ == "__main__":
    # Define a symbolic bivector. For example, let e12 be the
    ↪  bivector representing
    # a plane rotation in 2D or a component of a rotation in 3D.
    e12 = sp.symbols('e12')

    # Set the Lie algebra element as the bivector e12.
    bivector_expr = e12

    # Specify a rotation angle. For instance, pi/4.
    angle_val = sp.pi/4

    # Compute the rotor corresponding to the specified rotation.
    R = lie_group_to_ga(bivector_expr, angle_val, n_terms=15)
```

```python
print("Rotor R = exp(-1/2 * angle * B):")
sp.pprint(R)
print("\n" + "="*60 + "\n")

# Compute the commutator of e12 with itself (which should be
↪   zero).
comm = commutator(e12, e12)
print("Commutator [e12, e12]:")
sp.pprint(comm)
print("\n" + "="*60 + "\n")

# Compute the reverse (inverse) of the rotor.
R_rev = rotor_reverse(bivector_expr, angle_val, n_terms=15)
print("Rotor Reverse (Inverse) R_reverse = exp(+1/2 * angle *
↪   B):")
sp.pprint(R_rev)
print("\n" + "="*60 + "\n")

# Compose two rotations: for demonstration, compose R with its
↪   reverse.
R_composed = compose_rotors(R, R_rev)
print("Composition R * R_reverse (should yield identity):")
sp.pprint(sp.simplify(R_composed))
print("\n" + "="*60 + "\n")

# Compute the logarithm of the rotor (symbolic retrieval of the
↪   generator).
log_R = rotor_log(R)
print("Logarithm of the rotor, log(R):")
sp.pprint(log_R)
```

Chapter 44

Lie Algebras: Structure and Symmetry in GA

Algebraic Foundations of Lie Algebras in GA

Lie algebras serve as the mathematical backbone for describing infinitesimal generators of continuous transformations. Within the framework of geometric algebra (GA), the elements of a Lie algebra are often represented as bivectors or combinations of multivectors. The fundamental operation that encodes the structure of a Lie algebra is the Lie bracket, defined by

$$[X, Y] = XY - YX,$$

where X and Y belong to the vector space underlying the algebra. This antisymmetric operation encapsulates the non-commutative characteristics inherent to many physical symmetry operations. The algebraic perspective naturally organizes the generators and their relationships, with the structure constants emerging from the evaluation of the Lie bracket in a chosen basis. Such an approach unifies the geometric interpretation of rotations, boosts, and other symmetry transformations with their computational implementation.

Properties and Structural Aspects of the Lie Bracket

The Lie bracket in GA satisfies several critical properties that define the structure of the Lie algebra. Notably, the antisymmetry property

$$[X, Y] = -[Y, X],$$

ensures that every pair of elements interacts in a manner that preserves the signed structure of the commutator. Moreover, the Jacobi identity

$$[X, [Y, Z]] + [Y, [Z, X]] + [Z, [X, Y]] = 0$$

establishes a deep structural constraint, ensuring the closure of the algebra and dictating how the generators combine. The satisfaction of these identities implies that the set of GA elements, equipped with the Lie bracket, forms a well-defined Lie algebra. In applications to physics, the structure constants determined by these properties are directly linked to the conservation laws and invariant quantities that characterize physical systems.

Computational Techniques for Lie Algebra Operations

The implementation of Lie algebra operations in a computational setting is essential for bridging abstract theory and practical applications. A critical step is the calculation of the Lie bracket between two bivector elements. The following function computes the Lie bracket, encapsulating the intrinsic non-commutative behavior of GA elements. This function is designed to operate on symbolic representations of bivectors, leveraging the geometric product to compute the commutator.

```
def compute_lie_bracket(bivector1, bivector2):
    """
    Compute the Lie bracket (commutator) of two bivectors within the
    geometric algebra framework.

    The Lie bracket is defined by:
        [bivector1, bivector2] = bivector1 * bivector2 - bivector2 *
        ↪ bivector1
```

```
Parameters:
    bivector1 : symbolic representation of the first bivector.
    bivector2 : symbolic representation of the second bivector.

Returns:
    The commutator of the two bivectors, capturing the
    ↪ non-commutative structure
    intrinsic to the Lie algebra.
"""
return bivector1 * bivector2 - bivector2 * bivector1
```

The function above provides a computational tool that mirrors the algebraic definition of the Lie bracket. Such implementations facilitate symbolic manipulations and numerical experiments that probe the defining properties of Lie algebras in GA.

Representation of Physical Symmetries in GA

The mathematical formalism of Lie algebras is indispensable for the description of physical symmetries in both classical and quantum contexts. In GA, the symmetry operations of physical systems—ranging from rotations in three-dimensional space to more sophisticated transformations in relativistic settings—are generated by appropriate bivectors. The relationship between the infinitesimal generators and the finite transformation is expressed through the exponential mapping

$$\exp(B) = \sum_{k=0}^{\infty} \frac{B^k}{k!},$$

where B is a bivector acting as the generator. The non-commutative composition of these generators, as encoded in the Lie bracket, reflects the curvature and topology of the symmetry group manifold. This unified description allows for the extraction of conserved quantities and invariant features directly from the algebraic structure. The computational realization of these concepts in GA provides a fertile ground for the simulation and analysis of complex physical phenomena, ensuring that symmetry operations are implemented with both mathematical rigor and computational efficiency.

Python Code Snippet

```python
import math

class GAElement:
    """
    Represents an element of the even subalgebra of Cl(3,0),
    i.e. elements of the form: A = scalar + bivector.

    In this implementation, we represent the bivector part using
    three components corresponding to the basis bivectors:
      e23, e31, and e12.

    Thus, an element A is expressed as:
        A = a + b1*e23 + b2*e31 + b3*e12
    where 'a' is the scalar part.
    """
    def __init__(self, scalar=0.0, bivector=(0.0, 0.0, 0.0)):
        self.scalar = scalar
        # bivector is a tuple (b1, b2, b3)
        self.bivector = tuple(bivector)

    def __add__(self, other):
        if isinstance(other, GAElement):
            return GAElement(
                self.scalar + other.scalar,
                tuple(a + b for a, b in zip(self.bivector,
                 ↪    other.bivector))
            )
        else:
            # Assume 'other' is a scalar
            return GAElement(self.scalar + other, self.bivector)

    __radd__ = __add__

    def __sub__(self, other):
        if isinstance(other, GAElement):
            return GAElement(
                self.scalar - other.scalar,
                tuple(a - b for a, b in zip(self.bivector,
                 ↪    other.bivector))
            )
        else:
            return GAElement(self.scalar - other, self.bivector)

    def __rsub__(self, other):
        # When other is a scalar: other - self
        return GAElement(other - self.scalar, tuple(-b for b in
         ↪    self.bivector))

    def __mul__(self, other):
        if isinstance(other, (int, float)):
```

```python
        # Scalar multiplication
        return GAElement(
            self.scalar * other,
            tuple(b * other for b in self.bivector)
        )
    elif isinstance(other, GAElement):
        # Geometric product: (a + X) * (b + Y) = ab + aY + bX +
        ↪ X*Y
        #
        # For bivectors X and Y (with components X = (x1,x2,x3)
        ↪ and Y = (y1,y2,y3)),
        # we define their product as:
        #     X * Y = - (x1*y1 + x2*y2 + x3*y3) - (cross product
        ↪ of X and Y)
        #
        # The cross product (x1,x2,x3) x (y1,y2,y3) is computed
        ↪ as:
        #    (x2*y3 - x3*y2, x3*y1 - x1*y3, x1*y2 - x2*y1)
        #
        # Thus, the full product is:
        #
        #   scalar part: a*b - (x1*y1 + x2*y2 + x3*y3)
        #   bivector part: a*(y1,y2,y3) + b*(x1,x2,x3)
        #                  - (x2*y3 - x3*y2, x3*y1 - x1*y3,
        ↪ x1*y2 - x2*y1)
        a = self.scalar
        b = other.scalar
        x1, x2, x3 = self.bivector
        y1, y2, y3 = other.bivector

        scalar_part = a * b - (x1 * y1 + x2 * y2 + x3 * y3)

        # Contributions from multiplying scalar with bivector
        ↪ parts
        biv_from_scalars = (a * y1 + b * x1,
                            a * y2 + b * x2,
                            a * y3 + b * x3)
        # Bivector product: negative of the cross product
        cross_prod = (x2 * y3 - x3 * y2,
                      x3 * y1 - x1 * y3,
                      x1 * y2 - x2 * y1)
        bivector_part = (
            biv_from_scalars[0] - cross_prod[0],
            biv_from_scalars[1] - cross_prod[1],
            biv_from_scalars[2] - cross_prod[2],
        )

        return GAElement(scalar_part, bivector_part)
    else:
        raise TypeError("Multiplication with type {} not
        ↪ supported".format(type(other)))

__rmul__ = __mul__
```

```python
    def __neg__(self):
        return GAElement(-self.scalar, tuple(-b for b in
        ↪ self.bivector))

    def __str__(self):
        return f"{self.scalar} + {self.bivector[0]}*e23 +
        ↪ {self.bivector[1]}*e31 + {self.bivector[2]}*e12"

    def __repr__(self):
        return self.__str__()

    def norm(self):
        """
        For a pure bivector (with zero scalar part),
        the norm is given by: |B| = sqrt(b1^2 + b2^2 + b3^2).
        For general GAElements, this is not the full GA norm.
        Here, we assume use in the context of bivectors.
        """
        return math.sqrt(sum(b * b for b in self.bivector))

def compute_lie_bracket(X, Y):
    """
    Compute the Lie bracket (commutator) defined by:
        [X, Y] = X*Y - Y*X
    This function assumes that X and Y are GAElements (typically
    ↪ pure bivectors)
    representing Lie algebra generators in the geometric algebra
    ↪ framework.
    """
    return X * Y - Y * X

def exp_bivector(B, terms=20):
    """
    Compute the exponential of a bivector B using its closed form.

    For a simple bivector B (with B^2 = -|B|^2), we have:
        exp(B) = cos(|B|) + sin(|B|)/|B| * B.

    If the norm of B is zero, the function returns 1 (the identity
    ↪ element).
    """
    normB = B.norm()
    if normB == 0:
        return GAElement(1.0, (0.0, 0.0, 0.0))
    cos_term = math.cos(normB)
    sin_term = math.sin(normB)
    scaled_B = (sin_term / normB) * B
    return GAElement(cos_term, (0.0, 0.0, 0.0)) + scaled_B

def jacobi_identity(X, Y, Z):
    """
    Verify the Jacobi identity for Lie brackets:
```

```
    [X, [Y, Z]] + [Y, [Z, X]] + [Z, [X, Y]] = 0.
Returns the GAElement representing the sum, which should be
↪ (approximately) zero.
"""
    term1 = compute_lie_bracket(X, compute_lie_bracket(Y, Z))
    term2 = compute_lie_bracket(Y, compute_lie_bracket(Z, X))
    term3 = compute_lie_bracket(Z, compute_lie_bracket(X, Y))
    return term1 + term2 + term3

# Example usage demonstrating the key equations and algorithms
if __name__ == "__main__":
    # Define the basis bivectors for Cl(3,0):
    # e23, e31, and e12 are defined with unit coefficients.
    e23 = GAElement(0, (1.0, 0.0, 0.0))
    e31 = GAElement(0, (0.0, 1.0, 0.0))
    e12 = GAElement(0, (0.0, 0.0, 1.0))

    # Example 1: Compute Lie bracket between two bivector
    ↪ generators.
    # Let X = e12 and Y = e23.
    X = e12
    Y = e23
    bracket = compute_lie_bracket(X, Y)
    print("Lie bracket [X, Y] =", bracket)

    # Check antisymmetry: [X, Y] should be the negative of [Y, X].
    bracket_YX = compute_lie_bracket(Y, X)
    print("Lie bracket [Y, X] =", bracket_YX)
    print("Antisymmetry check, [X, Y] + [Y, X] =", bracket +
    ↪ bracket_YX)

    # Example 2: Compute the exponential of a bivector generator.
    # Using B = e12 as an example, the exponential yields a rotor.
    B = e12
    expB = exp_bivector(B)
    print("Exponential exp(B) =", expB)

    # Example 3: Verify the Jacobi identity with three bivector
    ↪ generators.
    # Choose X = e12, Y = e23, and Z = e31.
    Z = e31
    jacobi = jacobi_identity(X, Y, Z)
    print("Jacobi identity result [X,[Y,Z]] + [Y,[Z,X]] + [Z,[X,Y]]
    ↪ =", jacobi)
```

Chapter 45

Symmetry Transformations: Representing Physics with GA

Theoretical Foundations of Symmetry Transformations

In geometric algebra, symmetry operations are encapsulated by algebraic entities that unify and extend traditional representations of rotations, reflections, and other transformations. A rotation of a vector v is achieved via the sandwiching operation

$$v' = R v \tilde{R},$$

where R denotes a rotor and $\tilde{R}$ its reverse. The rotor is generated by the exponential mapping of a bivector which encodes the plane of rotation. For a rotation about an axis defined by a normalized bivector B (satisfying $B^2 = -1$), the rotor is expressed as

$$R = \exp\left(-\frac{\theta}{2} B\right).$$

This formulation integrates both the magnitude of the rotation and the geometric orientation of the rotational plane into a single compact operator. The algebraic structure ensures that the composition of such transformations respects the underlying symmetry group properties and preserves essential invariants such as magnitude and orientation.

Computational Methods for Generating Rotors

The computational implementation of symmetry transformations in geometric algebra involves the efficient evaluation of the exponential mapping of bivectors. When the bivector B is normalized, the infinite series expansion for the exponential truncates to a closed-form representation:

$$\exp\left(-\frac{\theta}{2}B\right) = \cos\left(\frac{\theta}{2}\right) - B\sin\left(\frac{\theta}{2}\right).$$

This closed-form expression is central to algorithmic strategies for applying rotations to physical vectors and multivectors. By normalizing the bivector, the complex interplay between trigonometric functions and the bivector structure is streamlined, yielding an operator that not only captures the rotation but does so while maintaining numerical stability. The rotor is then applied to physical quantities via the sandwiching operation, ensuring that the rotated entities inherit the correct geometric properties dictated by the symmetry of the system.

A concrete computational implementation involves encapsulating the rotor generation process in a dedicated function. The following Python function computes a rotor from a given bivector generator and rotation angle. The function first normalizes the input bivector, then applies the trigonometric mapping inherent in the exponential formulation, and finally returns the rotor as a new geometric algebra element.

```
def generate_rotor(bivector, angle):
    """
    Compute a rotor representing a rotation by the specified angle
    ↪ about
    the axis defined by the bivector in geometric algebra.

    The rotor R is computed via the exponential mapping:
      R = exp(-0.5 * angle * bivector_unit)
    where bivector_unit is the normalized version of the input
    ↪ bivector.
```

```
Parameters:
  bivector : a GAElement object representing the bivector
  ↪ generator.
  angle    : a float indicating the rotation angle in radians.

Returns:
  A GAElement object representing the rotor that encodes the
  ↪ rotation transformation.
"""
norm = bivector.norm()
if norm == 0:
    raise ValueError("Bivector must be non-zero to define a
    ↪ rotation axis.")
# Obtain the unit bivector corresponding to the rotation axis.
bivector_unit = (1.0 / norm) * bivector
half_angle = -0.5 * angle
cos_term = math.cos(half_angle)
sin_term = math.sin(half_angle)
# Assemble the rotor using the cosine and sine contributions.
rotor = GAElement(cos_term, (bivector_unit.bivector[0] *
↪ sin_term,
                             bivector_unit.bivector[1] *
                             ↪ sin_term,
                             bivector_unit.bivector[2] *
                             ↪ sin_term))
return rotor
```

This function exemplifies the translation of the theoretical rotor formulation into a practical algorithm. The rotor thus calculated serves as a robust computational tool for implementing symmetry transformations in various physical simulations.

Applications in Physical Systems

Symmetry operations are a cornerstone in the modeling of physical phenomena. In geometric algebra, the representation of these operations through rotors offers a unified and compact approach to simulate rotational dynamics and other transformation groups. For instance, given a vector v, its rotation by a specified angle is succinctly computed as

$$v' = Rv\tilde{R},$$

where R is generated from the bivector corresponding to the rotation plane. The intrinsic structure of the rotor preserves the magnitude and geometric integrity of v, ensuring that the physical interpretation remains consistent under transformation.

Beyond classical rotations, the rotor framework extends to more sophisticated symmetry transformations encountered in relativistic mechanics and quantum systems. The algebraic properties of rotors enable the consistent implementation of Lorentz boosts and other transformations encountered in the modeling of spacetime symmetries. In each case, the computational method hinges on the efficient evaluation of the exponential mapping and the subsequent application of the rotor via the sandwiching product. The result is a methodology that not only streamlines the theoretical derivations but also lends itself to high-performance numerical simulation, enabling the accurate reproduction of symmetry-driven phenomena across diverse areas of physics and computer science.

Python Code Snippet

```python
import math

class GAElement:
    """
    A simple representation of a Geometric Algebra element
    for rotations in 3D. In this context, an element is assumed
    to be of the form:

        GAElement = scalar + bivector,

    where the bivector is represented as a 3-tuple corresponding to
    the oriented plane components (analogous to the imaginary part
    in quaternions).
    """
    def __init__(self, scalar, bivector):
        self.scalar = scalar
        # Ensure bivector is stored as a tuple of three components.
        if isinstance(bivector, (list, tuple)):
            if len(bivector) != 3:
                raise ValueError("Bivector must have exactly three
                    components.")
            self.bivector = tuple(bivector)
        else:
            raise TypeError("Bivector must be a list or tuple of
                three numbers.")

    def __str__(self):
        return (f"{self.scalar:.6f} + "
                f"{self.bivector[0]:.6f}e12 + "
                f"{self.bivector[1]:.6f}e23 + "
                f"{self.bivector[2]:.6f}e31")

    def __repr__(self):
```

```python
        return f"GAElement(scalar={self.scalar}, 
          ↪ bivector={self.bivector})"

    def norm(self):
        """
        Computes the Euclidean norm of the GA element.
        For a rotor (scalar + bivector), this is analogous to the 
          ↪ quaternion norm.
        """
        return math.sqrt(self.scalar**2 + sum(comp**2 for comp in 
          ↪ self.bivector))

    def reverse(self):
        """
        Returns the reverse (or reversion) of the GA element.
        For rotor elements, the reverse flips the sign of the 
          ↪ bivector part.
        """
        return GAElement(self.scalar, tuple(-comp for comp in 
          ↪ self.bivector))

    def __mul__(self, other):
        """
        Implements the geometric product.
        If other is a scalar, performs scalar multiplication.
        If other is a GAElement, the product is defined analogously 
          ↪ to quaternion multiplication:

            (a0, a) * (b0, b) = (a0*b0 - a·b,  a0*b + b0*a + a×b)

        where a and b are the bivector parts (3-vectors), dot is the 
          ↪ scalar dot product,
        and × is the standard cross product.
        """
        if isinstance(other, (int, float)):
            return GAElement(self.scalar * other,
                             tuple(comp * other for comp in 
                               ↪ self.bivector))
        elif isinstance(other, GAElement):
            s1 = self.scalar
            s2 = other.scalar
            v1 = self.bivector
            v2 = other.bivector
            dot = v1[0]*v2[0] + v1[1]*v2[1] + v1[2]*v2[2]
            # Cross product: v1 x v2
            cross = (v1[1]*v2[2] - v1[2]*v2[1],
                     v1[2]*v2[0] - v1[0]*v2[2],
                     v1[0]*v2[1] - v1[1]*v2[0])
            new_scalar = s1 * s2 - dot
            new_bivector = (s1*v2[0] + s2*v1[0] + cross[0],
                            s1*v2[1] + s2*v1[1] + cross[1],
                            s1*v2[2] + s2*v1[2] + cross[2])
            return GAElement(new_scalar, new_bivector)
```

```python
        else:
            return NotImplemented

    def __rmul__(self, other):
        # Allows scalar multiplication from the left.
        return self.__mul__(other)

    def __add__(self, other):
        if isinstance(other, GAElement):
            new_scalar = self.scalar + other.scalar
            new_bivector = tuple(a + b for a, b in
                ↪  zip(self.bivector, other.bivector))
            return GAElement(new_scalar, new_bivector)
        else:
            return NotImplemented

def generate_rotor(bivector, angle):
    """
    Compute a rotor representing a rotation by the specified angle
    ↪  about
    the axis defined by the bivector in geometric algebra.

    The rotor R is computed via the exponential mapping:
      R = exp(-0.5 * angle * bivector_unit)
    where bivector_unit is the normalized version of the input
    ↪  bivector.

    Parameters:
      bivector : a GAElement object representing the bivector
      ↪  generator.
      angle    : a float indicating the rotation angle in radians.

    Returns:
      A GAElement object representing the rotor that encodes the
      ↪  rotation transformation.
    """
    norm_b = bivector.norm()
    if norm_b == 0:
        raise ValueError("Bivector must be non-zero to define a
            ↪  rotation axis.")
    # Obtain the unit bivector corresponding to the rotation axis.
    bivector_unit = (1.0 / norm_b) * bivector
    half_angle = -0.5 * angle
    cos_term = math.cos(half_angle)
    sin_term = math.sin(half_angle)
    # Assemble the rotor using the cosine and sine contributions.
    rotor = GAElement(cos_term, tuple(sin_term * comp for comp in
        ↪  bivector_unit.bivector))
    return rotor

def rotate_vector(vector, rotor):
    """
    Rotate a vector using the rotor.
```

```
    The vector is embedded as a GAElement with zero scalar part.
    The rotation is performed by the sandwich product:
        v' = R * v * R.reverse()

    Parameters:
      vector : a tuple of 3 floats representing the vector to be
      ↪  rotated.
      rotor  : a GAElement object representing a rotor.

    Returns:
      A tuple of 3 floats representing the rotated vector.
    """
    # Represent the vector as a pure GAElement (zero scalar, vector
    ↪  as bivector components)
    v = GAElement(0, vector)
    # Apply the sandwich product to rotate.
    rotated = rotor * v * rotor.reverse()
    # In a proper rotation, the result should have zero scalar part.
    return rotated.bivector

if __name__ == "__main__":
    # Example: Rotate the vector (1, 0, 0) by 90 degrees about the
    ↪  z-axis.

    # Define the bivector representing the plane of rotation.
    # For rotation about the z-axis in a 3D space, the corresponding
    ↪  bivector can be
    # represented by (0, 0, 1) (analogous to the imaginary unit in
    ↪  quaternions).
    bivector = GAElement(0, (0, 0, 1))

    # Define rotation angle in radians (90 degrees).
    angle = math.pi / 2  # 90 degrees

    # Generate the rotor using the exponential mapping method.
    rotor = generate_rotor(bivector, angle)
    print("Generated Rotor:")
    print(rotor)

    # The rotor encodes the rotation. To apply it to a vector,
    # use the sandwich product v' = R * v * R.reverse().
    vector = (1, 0, 0)  # Initial vector along the x-axis.
    rotated_vector = rotate_vector(vector, rotor)

    print("\nOriginal Vector:")
    print(vector)
    print("\nRotated Vector:")
    print(rotated_vector)
```

Chapter 46

Conservation Laws in GA: Deriving Invariants

Algebraic Foundations and Invariant Structures

Geometric algebra (GA) provides a framework in which the synthesis of scalars, vectors, and higher-grade elements leads to intrinsic conservation properties. The underlying algebraic structure allows the formation of invariants by considering the geometric product and the reversion operation. For any multivector M, the product

$$I = M \tilde{M}$$

yields a scalar quantity that remains unchanged under GA transformations. When M is expressed as a sum of a scalar component and a bivector part, such that

$$M = \alpha + A,$$

the invariant quantity defined by

$$\|M\|^2 = M \tilde{M} = \alpha^2 + \langle A^2 \rangle$$

encapsulates conservation laws analogous to those found in classical physics. This construction is of paramount importance when identifying conserved quantities such as energy or momentum within physical systems. The property

$$R\tilde{R} = 1,$$

which holds for rotors R, further reinforces the idea that GA operations preserve inherent geometric invariants.

Derivation of Invariants from Geometric Structures

The derivation of conservation laws in GA is intrinsically linked to the algebra's ability to encode symmetry properties. When a physical system is represented by a GA element, the inherent bilinear forms naturally lead to conserved scalar quantities. For example, consider a GA element representing a physical observable:

$$E = a + \mathbf{B},$$

where a is a scalar and $\mathbf{B}$ is a bivector corresponding to angular momentum or other rotational characteristics. The product

$$E\tilde{E} = a^2 + \|\mathbf{B}\|^2$$

remains invariant under transformations that preserve the multivector's structure. This derivation rests on the fact that operations such as reversion and Clifford conjugation within GA maintain the scalar part of the geometric product, yielding invariants that reflect conserved physical quantities. The capability to decompose a multivector into contributing grades and recombine them through the geometric product enables the extraction of invariants that underpin the conservation laws observed in various physical contexts.

Computational Approaches to GA Invariants

The computational implementation of conservation laws within the GA framework benefits from the algebra's algorithmic clarity. In simulations involving the time evolution of physical systems, verifying that invariants remain constant serves as a diagnostic for numerical stability and physical correctness. The invariant associated with any GA element can be computed by evaluating the sum of the square of its scalar component and the squared norm of its bivector parts. A typical function to perform this computation is provided below.

```
def compute_conservation_invariant(ga_element):
    """
    Compute the invariant associated with a geometric algebra
    ↪ element.

    For a GA element represented by a scalar part and a bivector
    ↪ part,
    the invariant quantity is given by the sum of the square of the
    ↪ scalar
    and the squared Euclidean norm of the bivector components:

        invariant = (scalar)**2 + (||bivector||)**2,

    where (||bivector||) is computed as the square root of the sum
    ↪ of the squares
    of the bivector components.

    Parameters:
        ga_element : An object with attributes 'scalar' (float) and
        ↪ 'bivector' (tuple of floats)

    Returns:
        A float representing the invariant of the GA element, which
        ↪ remains constant
        under GA transformations.
    """
    bivector_norm = sum(component**2 for component in
    ↪ ga_element.bivector)**0.5
    return ga_element.scalar**2 + bivector_norm**2
```

The function computes the conserved quantity by first determining the Euclidean norm of the bivector components and then combining it with the square of the scalar part. In a computational simulation, the constant value returned by this function serves as an indicator that the numerical procedure respects the invariance dictated by GA. This invariant plays a central role in verifying the conservation of energy, momentum, and other physical quantities during time evolution.

The interplay between the algebraic derivation of invariants and its computational implementation exemplifies how GA provides a unifying language for both theoretical analysis and high-performance numerical simulation in physics.

Python Code Snippet

```python
import math

class GAElement:
    """
    Class representing an element in Geometric Algebra (GA) with a
    ↪ scalar part and a bivector part.
    A GA element is expressed as:
        M = scalar + bivector
    where 'scalar' is a float and 'bivector' is a list (or tuple) of
    ↪ floats.
    """
    def __init__(self, scalar, bivector):
        self.scalar = scalar
        # Ensure that the bivector is stored as a list of floats
        self.bivector = list(bivector)

    def __repr__(self):
        return f"GAElement(scalar={self.scalar},
        ↪ bivector={self.bivector})"

    def reversion(self):
        """
        Compute the reversion (or reversal) of the GA element.
        For an element M = scalar + bivector, the reversion is
        ↪ defined as:
            M_tilde = scalar - bivector
        This operation reverses the sign of all bivector components.
        """
        negated_bivector = [-comp for comp in self.bivector]
        return GAElement(self.scalar, negated_bivector)

    def geometric_product(self, other):
        """
        Compute the geometric product of two GA elements.

        For two elements M = a + A and N = b + B, where A and B are
        ↪ bivector parts,
        a simplified version of the geometric product is
        ↪ implemented.

        The product is given by:
            M * N = ab + aB + bA + (A * B)

        In our simplified model:
          - The bivector product A * B is taken as the negative dot
          ↪ product: - (A · B)
          - The cross terms aB and bA are computed component-wise.

        Note: This implementation is for demonstration purposes and
        ↪ is limited to GA
```

329

*elements consisting of a scalar and a bivector. More complex
↪ multivector products
require a complete treatment of grade mixing.*
"""
```
# Multiply scalar parts
scalar_part = self.scalar * other.scalar

# Cross terms: scalar multiplying the bivector components
bivector_part = [self.scalar * b + other.scalar * a
                 for a, b in zip(self.bivector,
                 ↪  other.bivector)]

# Bivector multiplication contribution (simplified as
↪ negative dot product)
bivector_dot = sum(a * b for a, b in zip(self.bivector,
↪  other.bivector))
scalar_bivector_part = -bivector_dot

# Combine scalar contributions
total_scalar = scalar_part + scalar_bivector_part

# Construct and return the resulting GA element.
# Note: In a full implementation, additional terms (e.g.,
↪ higher grade parts) would be considered.
return GAElement(total_scalar, bivector_part)

def norm_squared(self):
    """
```
*Computes the invariant norm squared of the GA element,
↪ defined as:*

*$||M||^2 = M * M\_tilde = (scalar)^2 + (Euclidean\ norm\ of$
↪ $bivector)^2$*

*This invariant remains constant under GA transformations and
↪ encapsulates
conservation laws analogous to energy or momentum in
↪ physical systems.*
"""
```
    bivector_norm_sq = sum(comp ** 2 for comp in self.bivector)
    return self.scalar ** 2 + bivector_norm_sq

def compute_conservation_invariant(ga_element):
    """
```
Compute the conservation invariant for a GA element.

*For a GA element M represented as (scalar + bivector), the
↪ invariant is:*

$invariant = (scalar)^2 + (||bivector||)^2$,

*where $||bivector||$ is the Euclidean norm computed as the square
↪ root of the sum of*

```
        the squares of the bivector components.

        Parameters:
            ga_element : An instance of GAElement

        Returns:
            A float representing the invariant of the GA element.
        """
        bivector_norm = math.sqrt(sum(component ** 2 for component in
        ↪    ga_element.bivector))
        return ga_element.scalar ** 2 + bivector_norm ** 2

# Example Usage and Test Cases
if __name__ == "__main__":
    # Create a GA element M = alpha + A, where alpha is a scalar and
    ↪    A is a bivector.
    M = GAElement(3.0, [1.0, 2.0, 3.0])
    print("Original GA Element M:", M)

    # Compute the reversion of M (i.e., M_tilde = scalar - bivector)
    M_tilde = M.reversion()
    print("Reversion of M:", M_tilde)

    # Compute the invariant via the norm_squared() method: M *
    ↪    M_tilde = (scalar)^2 + ||A||^2
    invariant_method = M.norm_squared()
    print("Invariant from norm_squared method:", invariant_method)

    # Alternatively, compute the invariant using the dedicated
    ↪    function
    invariant_function = compute_conservation_invariant(M)
    print("Invariant from compute_conservation_invariant function:",
    ↪    invariant_function)

    # Demonstrate the rotor property: For a rotor R, we require that
    ↪    R * R_tilde = 1.
    # Let R = cos(theta/2) + sin(theta/2)*n where n is a unit
    ↪    bivector.
    theta = math.pi / 4   # 45 degrees rotation
    cos_half = math.cos(theta / 2)
    sin_half = math.sin(theta / 2)
    # Example rotor with bivector components chosen for a 2D
    ↪    rotation (e.g., [sin(theta/2), 0])
    R = GAElement(cos_half, [sin_half, 0.0])
    print("Rotor R:", R)
    rotor_invariant = R.norm_squared()
    print("Rotor invariant (should be close to 1):",
    ↪    rotor_invariant)
```

Chapter 47

Noether's Theorem in GA: Mathematical Insights

Algebraic Reformulation of Noether's Theorem in Geometric Algebra

Within the framework of geometric algebra (GA), the classical statement of Noether's theorem is reinterpreted through the intrinsic algebraic structure of multivectors. In this formulation, continuous symmetries are encoded in the operations of GA and the associated conserved quantities are derived from invariant scalar products. A GA element is typically expressed in the form

$$M = \alpha + A,$$

where α denotes the scalar component and A represents higher-grade elements such as bivectors. The invariant

$$I = M\tilde{M},$$

with $\tilde{M}$ being the reversion of M, provides a natural generalization of the traditional conserved quantity. The invariance of I under symmetry transformations is a direct consequence of the underlying properties of the geometric product, which remains unaltered when the multivector undergoes rotations or other continuous transformations expressed by rotors.

Symmetry Transformations and GA Invariants

The reformulation of Noether's theorem in a GA context emphasizes the role of symmetry in determining conservation laws. In GA, a rotor R is defined such that

$$R\tilde{R} = 1,$$

which ensures that the transformation of a GA element preserves its inherent invariant structure. When a system is subjected to a continuous group of GA transformations, the corresponding Lagrangian, represented in terms of GA elements, exhibits invariance that directly leads to the formulation of conserved currents. The reversion operation, analogous to a complex conjugation in certain contexts, ensures that products such as $M\tilde{M}$ yield real-valued invariants. This approach not only streamlines the derivation of conservation laws but also provides a unified algebraic language that links symmetry and invariance in both mathematical and physical terms.

Computational Implementation of GA Noether Invariants

The translation of the abstract algebraic invariants into a computational algorithm is a powerful demonstration of the GA approach. The invariant associated with a GA element is computed by taking the geometric product of the element with its reversion. Such a computation is pivotal in numerical simulations of physical systems where conservation laws serve as stringent consistency checks.

```
def compute_noether_invariant(ga_element):
    """
    Compute the Noether invariant associated with a GA element.
    
    In the geometric algebra framework, a GA element M is composed
      as
    M = scalar + bivector, and its Noether invariant is defined by
      the
    geometric product with its reversion: invariant = M * ~M.
    This function assumes that the geometric product operation and
      the
    reversion method are defined for the input ga_element.
```

```
Parameters:
    ga_element : An object representing a GA element with
    ↪ methods
                'reversion()' and 'geometric_product(other)'.

Returns:
    A float representing the scalar component of the invariant,
    which remains constant under GA symmetric transformations.
"""
reversed_element = ga_element.reversion()
product = ga_element.geometric_product(reversed_element)
return product.scalar
```

The function presented above encapsulates the computational procedure for obtaining the invariant. The product $M \tilde{M}$ is computed by first obtaining the reversion of the GA element and then applying the geometric product. The scalar part of the resulting multivector is extracted as the conserved quantity. This numeric implementation reinforces the mathematical insights that Noether's theorem, when expressed in the language of GA, naturally yields conserved invariants that are robust under symmetry transformations.

Python Code Snippet

```
import math

class GAElement:
    """
    A simple representation of a Geometric Algebra (GA) element
    in a 2D Euclidean space, expressed as a sum of a scalar and
    a bivector component (associated with the basis element e12).

    A GA element is written as:
        M = scalar + bivector * e12
    """
    def __init__(self, scalar=0.0, bivector=0.0):
        self.scalar = scalar
        self.bivector = bivector

    def reversion(self):
        """
        Compute the reversion (or reversal) of the GA element.
        For an element M = scalar + bivector*e12, the reversion is:
            ~M = scalar - bivector*e12
        """
        return GAElement(self.scalar, -self.bivector)
```

```python
    def geometric_product(self, other):
        """
        Compute the geometric product of two GA elements.

        Given two GA elements:
            M = a + b*e12
            N = c + d*e12
        the product is defined by:
            M * N = (a*c - b*d) + (a*d + b*c) * e12

        Here, we use the property that e12*e12 = -1 (Euclidean
        ↪ metric).
        """
        a, b = self.scalar, self.bivector
        c, d = other.scalar, other.bivector
        new_scalar = a * c - b * d
        new_bivector = a * d + b * c
        return GAElement(new_scalar, new_bivector)

    def __mul__(self, other):
        """
        Overload the multiplication operator to perform the
        geometric product between GA elements.
        """
        return self.geometric_product(other)

    def __str__(self):
        """
        Return a human-readable string representation of the GA
        ↪ element.
        """
        return f"{self.scalar} + {self.bivector}e12"

def compute_noether_invariant(ga_element):
    """
    Compute the Noether invariant associated with a GA element.

    In the GA formulation of Noether's theorem, the invariant is
    ↪ obtained from:
        I = M * ~M,
    where ~M denotes the reversion of M. The scalar part of I
    represents the conserved quantity under GA symmetry
    ↪ transformations.
    """
    reversed_element = ga_element.reversion()
    product = ga_element * reversed_element
    return product.scalar

def create_rotor(theta):
    """
    Create a rotor for a given rotation angle theta (in radians).
```

335

```
In geometric algebra, a rotor R is defined as:
    R = cos(theta/2) + sin(theta/2) * e12,
satisfying the normalization condition:
    R * ~R = 1.
"""
return GAElement(math.cos(theta/2), math.sin(theta/2))

def rotate_ga_element(ga_element, theta):
    """
    Rotate a GA element using the rotor corresponding to the
    ↪ rotation
    angle theta.

    The rotated element M' is computed via the sandwiching
    transformation:
        M' = R * M * ~R,
    where R is the rotor generated for the angle theta.
    """
    R = create_rotor(theta)
    rotated = R * ga_element * R.reversion()
    return rotated

# Example usage demonstrating the computation of the Noether
↪ invariant
if __name__ == "__main__":
    # Define a GA element M = 3 + 4e12
    M = GAElement(3, 4)
    invariant_original = compute_noether_invariant(M)

    print("Original GA Element M:", M)
    print("Noether Invariant (M * ~M):", invariant_original)

    # Rotate the element by 60 degrees (/3 radians)
    theta = math.pi / 3
    M_rotated = rotate_ga_element(M, theta)
    invariant_rotated = compute_noether_invariant(M_rotated)

    print("\nAfter Rotation by 60 Degrees:")
    print("Rotated GA Element M':", M_rotated)
    print("Noether Invariant (M' * ~M'):", invariant_rotated)

    # Verify the invariance under rotation
    if math.isclose(invariant_original, invariant_rotated,
    ↪ rel_tol=1e-9):
        print("\nInvariant is preserved under GA rotation.")
    else:
        print("\nInvariant is NOT preserved. Check the
        ↪ implementation!")
```

Chapter 48

Gauge Theory Foundations in GA: A Unified Treatment

Mathematical Formulation of GA Gauge Fields

In the geometric algebra framework, gauge fields are represented as multivector functions defined over spacetime. The conventional gauge potential, traditionally denoted as $A(x)$, is reformulated as a GA element that encapsulates both vector and bivector components. This representation exploits the capability of the geometric product to combine different grades into a unified algebraic structure. A key element of gauge theories is the local transformation dictated by a rotor R, which satisfies

$$R\tilde{R} = 1,$$

where $\tilde{R}$ denotes the reversion of R. The transformation of the gauge field is expressed by

$$A'(x) = R\,A(x)\,\tilde{R} + (\nabla R)\,\tilde{R},$$

with ∇R representing the derivative of the rotor with respect to the spacetime coordinates. This formulation naturally integrates the connection and derivative operators into a cohesive algebraic machinery.

Representation of Gauge Connections Using Geometric Algebra

Gauge connections in GA are embedded within the multivector structure of the algebra, where the derivative operator ∇ acts as a generalized vector derivative. The covariant derivative is defined by combining the gauge potential with the differentiation operator, yielding an expression of the form

$$D = \nabla + A(x).$$

The curvature or field strength tensor, obtained from the commutator of covariant derivatives, takes the form

$$F(x) = \nabla A(x) + A(x)^2.$$

The operation $A(x)^2$, computed using the geometric product, naturally includes contributions from both the vector and bivector parts of $A(x)$. The invariance of $F(x)$ under the local gauge transformation

$$A'(x) = R\,A(x)\,\tilde{R} + (\nabla R)\,\tilde{R},$$

demonstrates the self-consistency of the GA approach, wherein all elements remain within the same algebraic framework.

Computational Implementation of Gauge Transformations in GA

The computational implementation of gauge transformations is streamlined by the unified treatment intrinsic to geometric algebra. The transformation of a gauge field under a rotor R involves the computation of the sandwich product and an additional derivative term. The transformation is given by

$$A'(x) = R\,A(x)\,\tilde{R} + (\nabla R)\,\tilde{R}.$$

This operation is central to numerical simulations, where efficient algorithms must perform these algebraic manipulations reliably. The following Python function illustrates an algorithmic implementation of the gauge transformation. It encapsulates the necessary steps by applying the rotor transformation to the gauge field and incorporating the derivative contribution arising from a varying rotor.

```
def transform_gauge_field(ga_field, rotor, derivative_rotor):
    """
    Calculate the gauge-transformed GA field.

    This function applies the gauge transformation to a GA-valued
    ↪ field.
    The transformation is defined by:
        A' = R * A * ~R + (dR) * ~R,
    where R is the rotor representing the gauge change, ~R is its
    ↪ reversion,
    and dR is the derivative of the rotor field with respect to the
    ↪ spacetime coordinate.

    Parameters:
        ga_field: GA element representing the gauge field A.
        rotor: GA element acting as the gauge transformation rotor
        ↪ R.
        derivative_rotor: GA element representing the derivative of
        ↪ the rotor dR.

    Returns:
        GA element corresponding to the transformed gauge field A'.
    """
    return rotor * ga_field * rotor.reversion() + derivative_rotor *
    ↪ rotor.reversion()
```

This function demonstrates the application of a rotor to a given GA gauge field. The operation defines a coherent computational pathway for implementing local gauge transformations, where the geometric product and the reversion operation preserve the intrinsic symmetry properties of the field. The method integrates the algebraic structure of GA with efficient numerical strategies, offering a unified approach to simulate gauge-invariant physical phenomena.

Python Code Snippet

```
# Import necessary libraries (here we use numpy for numerical
↪ operations)
import numpy as np

class GAElement:
    """
    A rudimentary class representing a multivector (GA element)
    ↪ using a dictionary.
    Keys represent the grade signature (for example, (0,) for
    ↪ scalars) and values are the coefficients.
```

```
Note: This is a minimal demonstration. In practice, a dedicated
↪ geometric algebra library
(such as clifford or galgebra) should be used for full
↪ functionality.
"""
def __init__(self, components):
    # components: dict mapping grade identifiers (tuples) to
    ↪ their coefficients.
    # Example: {(0,): 1.0} represents the scalar 1.0.
    self.components = components

def __add__(self, other):
    result = self.components.copy()
    for grade, coef in other.components.items():
        result[grade] = result.get(grade, 0) + coef
    return GAElement(result)

def __sub__(self, other):
    result = self.components.copy()
    for grade, coef in other.components.items():
        result[grade] = result.get(grade, 0) - coef
    return GAElement(result)

def __mul__(self, other):
    """
    Naively implements the geometric product.
    For this demonstration, we assume multiplication only
    ↪ involves scalar parts (grade (0,)).
    A full implementation would handle grade mixing and blade
    ↪ multiplication.
    """
    if isinstance(other, GAElement):
        res = {}
        # If both elements have a scalar part, multiply them.
        if (0,) in self.components and (0,) in other.components:
            res[(0,)] = self.components[(0,)] *
                ↪ other.components[(0,)]
        # More complex grade interactions would be implemented
        ↪ here.
        return GAElement(res)
    else:
        raise ValueError("Multiplication with type {} not
            ↪ supported".format(type(other)))

def __pow__(self, power):
    # Exponentiation is only defined for scalar elements in this
    ↪ simple example.
    if (0,) in self.components:
        return GAElement({(0,): self.components[(0,)] ** power})
    else:
        raise ValueError("Exponentiation not implemented for
            ↪ non-scalar GA elements.")
```

340

```python
    def reversion(self):
        """
        Returns the reversion of the GA element.
        In a full implementation, this operation reverses the order
        ↪    of the basis vectors
        in each blade. Here, it is assumed the element is
        ↪    self-reversible.
        """
        return self

    def __repr__(self):
        return f"GAElement({self.components})"


def transform_gauge_field(ga_field, rotor, derivative_rotor):
    """
    Compute the gauge-transformed field in geometric algebra.

    The transformation is defined as:
        A'(x) = R * A(x) * ~R + (R) * ~R,
    where:
        - R is the rotor effecting the local gauge transformation.
        - ~R is the reversion of the rotor.
        - A(x) is the original GA gauge field.
        - R represents the derivative (gradient) of the rotor.

    Parameters:
        ga_field: GAElement representing the gauge field A(x).
        rotor: GAElement representing the gauge transformation rotor
        ↪    R.
        derivative_rotor: GAElement representing the derivative of R
        ↪    with respect to spacetime.

    Returns:
        GAElement corresponding to the transformed gauge field A'(x).
    """
    return rotor * ga_field * rotor.reversion() + derivative_rotor *
    ↪    rotor.reversion()


def covariant_derivative(gradient, gauge_field):
    """
    Compute the covariant derivative D =  + A(x),
    where:
        -  is the derivative operator (represented as a GAElement).
        - A(x) is the gauge field.

    Parameters:
        gradient: GAElement representing the derivative operator .
        gauge_field: GAElement representing the gauge field A(x).

    Returns:
        GAElement representing the covariant derivative D.
```

```
    """
    return gradient + gauge_field


def curvature_field(gradient, gauge_field):
    """
    Compute the curvature (field strength) F(x) in geometric
    ↪ algebra:
        F(x) = A(x) + A(x)^2.

    This captures both the derivative contribution and the
    ↪ self-interaction term computed via the geometric product.

    Parameters:
        gradient: GAElement representing the derivative operator .
        gauge_field: GAElement representing the gauge field A(x).

    Returns:
        GAElement representing the curvature F(x).
    """
    # Here we use the simplified geometric product for
    ↪ demonstration.
    return gradient * gauge_field + gauge_field * gauge_field


# Example usage demonstrating the application of the above
↪ operations:

# Define a simple gauge field A(x) as a scalar GAElement (for
↪ demonstration, we use a scalar value 2.0)
A = GAElement({(0,): 2.0})

# Define a rotor R. For a proper gauge transformation, R must
↪ satisfy R * ~R = 1.
# Here, we choose a trivial rotor with scalar value 1.0.
R = GAElement({(0,): 1.0})

# Define the derivative of the rotor dR. In this demo, we model it
↪ as a scalar with value 0.1.
dR = GAElement({(0,): 0.1})

# Define a gradient operator as a placeholder GAElement (scalar
↪ value 0.5 in this simplified example).
grad = GAElement({(0,): 0.5})

# Compute the transformed gauge field A'(x) using the gauge
↪ transformation.
A_transformed = transform_gauge_field(A, R, dR)
print("Transformed Gauge Field A':", A_transformed)

# Compute the covariant derivative D =  + A.
D = covariant_derivative(grad, A)
print("Covariant Derivative D:", D)
```

```
# Compute the curvature field F(x) = A + A^2.
F = curvature_field(grad, A)
print("Curvature Field F:", F)
```

Chapter 49

Spin Representations: Computation with GA

Mathematical Formalism of Spin Representations

Within the geometric algebra framework, spin is represented by elements of the even subalgebra that naturally encode rotations and quantum properties. In this formulation, a spinor is interpreted as a rotor and is constructed from a bivector that specifies the plane of rotation. The rotor is defined by the exponential mapping

$$R = \exp\left(-\frac{\theta}{2}B\right) = \cos\frac{\theta}{2} - B\sin\frac{\theta}{2},$$

where B is a normalized bivector satisfying $B^2 = -1$, and θ is the rotation angle. This representation unifies the treatment of rotation operators and spin states, as the action on a multivector M is performed via the sandwich operation

$$M' = R M \tilde{R},$$

with $\tilde{R}$ denoting the reversion of R. The formalism provides an algebraic structure that captures both the magnitude and phase information essential in quantum mechanics.

Computational Techniques for Spin Transformations

A pivotal computational technique in the geometric algebra representation of spin involves the explicit construction of rotors from bivector generators. The above formulation leads to efficient algorithms where the rotor acts as a bridge between abstract algebraic operations and numerical implementations. The rotor, when computed correctly, guarantees that the rotation transformation is norm-preserving, as ensured by the relation

$$R\tilde{R} = 1.$$

A computational routine may encapsulate this process using a function that takes a bivector and a rotation angle and returns the corresponding rotor. The following Python function encapsulates the exponential mapping for a normalized bivector:

```python
def compute_rotor(bivector, theta):
    """
    Compute the rotor corresponding to a rotation defined by a
      bivector and an angle.

    The rotor is computed as:
        R = cos(theta/2) - B * sin(theta/2),
    where B is a normalized bivector that satisfies B^2 = -1.

    Parameters:
        bivector: An object representing the normalized bivector,
          according to GA conventions.
        theta:    The rotation angle in radians.

    Returns:
        An object representing the rotor in geometric algebra.
    """
    import math
    cos_half = math.cos(theta / 2)
    sin_half = math.sin(theta / 2)
    R = cos_half - bivector * sin_half
    return R
```

This function computes the rotor by evaluating the cosine and sine of half the rotation angle and then combining these with the input bivector. The negative sign preceding the bivector term adheres to the convention for rotor construction in geometric algebra.

Spinor Dynamics and Computational Operations

The dynamics of spinor transformations are governed by the sequential application of rotors. Successive rotations are computed through the geometric product of the individual rotors:

$$R_{\text{total}} = R_2 \, R_1,$$

where each rotor R_1 and R_2 is constructed via the exponential map from its associated bivector generator. This property simplifies the composition of rotations and allows for the direct computation of spinor evolution under various quantum operations. The geometric product inherently encapsulates non-commutative behavior, which is critical for accurately modeling the underlying physics of spin systems.

The computational scheme relies on efficient algorithms for the geometric product and reversion operations. These algorithms form the backbone of numerical simulations that involve spinor manipulation, and they provide a unified treatment of the algebraic operations required. The combined use of analytical formulations and algorithmic implementations renders the GA framework particularly powerful for simulating spin phenomena in quantum mechanics.

Python Code Snippet

```
import math

class GA2D:
    """
    A minimal implementation of 2D Geometric Algebra elements.
    A GA2D element is represented as:
        A = scalar + e1*e1_coef + e2*e2_coef + e12*e12_coef,
    where e12 = e1e2 is the bivector and satisfies e12^2 = -1.
    """
    def __init__(self, scalar=0.0, e1=0.0, e2=0.0, e12=0.0):
        self.scalar = scalar
        self.e1 = e1
        self.e2 = e2
        self.e12 = e12

    def __add__(self, other):
        return GA2D(self.scalar + other.scalar,
```

```python
                    self.e1 + other.e1,
                    self.e2 + other.e2,
                    self.e12 + other.e12)

    def __sub__(self, other):
        return GA2D(self.scalar - other.scalar,
                    self.e1 - other.e1,
                    self.e2 - other.e2,
                    self.e12 - other.e12)

    def __mul__(self, other):
        # Scalar multiplication
        if isinstance(other, (int, float)):
            return GA2D(self.scalar * other,
                        self.e1 * other,
                        self.e2 * other,
                        self.e12 * other)
        # Geometric product for GA2D elements
        if isinstance(other, GA2D):
            a, b, c, d = self.scalar, self.e1, self.e2, self.e12
            p, q, r, s = other.scalar, other.e1, other.e2, other.e12
            new_scalar = a * p + b * q + c * r - d * s
            new_e1     = a * q + b * p - c * s + d * r
            new_e2     = a * r + b * s + c * p - d * q
            new_e12    = a * s + b * r - c * q + d * p
            return GA2D(new_scalar, new_e1, new_e2, new_e12)
        raise TypeError("Multiplication with type {} not
         ↪ supported".format(type(other)))

    def __rmul__(self, other):
        # To support scalar multiplication from left.
        return self.__mul__(other)

    def reverse(self):
        """
        Reversion (or reversal) flips the sign of bivector
         ↪ components.
        For a general multivector: A^~ = scalar + e1 + e2 - e12.
        """
        return GA2D(self.scalar, self.e1, self.e2, -self.e12)

    def __str__(self):
        parts = []
        if abs(self.scalar) > 1e-9:
            parts.append(f"{self.scalar:.4f}")
        if abs(self.e1) > 1e-9:
            parts.append(f"{self.e1:.4f}e1")
        if abs(self.e2) > 1e-9:
            parts.append(f"{self.e2:.4f}e2")
        if abs(self.e12) > 1e-9:
            parts.append(f"{self.e12:.4f}e12")
        if not parts:
            return "0"
```

```python
        return " + ".join(parts)

def compute_rotor(bivector, theta):
    """
    Compute the rotor corresponding to a rotation defined by a
    ↪ normalized bivector and an angle theta.

    The rotor is obtained via the exponential mapping:
        R = exp(- (theta/2) * B) = cos(theta/2) - B*sin(theta/2),
    where 'bivector' is assumed to be a GA2D element with only the
    ↪ e12 component nonzero
    (i.e., B = 0 + 0*e1 + 0*e2 + 1*e12, such that B^2 = -1).
    """
    cos_half = math.cos(theta / 2)
    sin_half = math.sin(theta / 2)
    # Construct rotor: R = cos(theta/2) - bivector * sin(theta/2)
    rotor = GA2D(cos_half, 0, 0, 0) - bivector * sin_half
    return rotor

def apply_rotor(multivector, rotor):
    """
    Apply a rotor to a multivector using the sandwich product:
        M' = rotor * multivector * (rotor.reverse())
    This operation rotates the multivector in the plane defined by
    ↪ the bivector generator.
    """
    return rotor * multivector * rotor.reverse()

def combine_rotors(rotor1, rotor2):
    """
    Combine two rotors using the geometric product.
    The combined rotor represents the successive application of the
    ↪ individual rotations.
        R_total = rotor2 * rotor1.
    """
    return rotor2 * rotor1

# Example demonstration of spin representations and transformations
if __name__ == "__main__":
    # Define a normalized bivector in 2D GA (unit bivector for
    ↪ rotation)
    B = GA2D(0, 0, 0, 1)  # Represents the unit bivector e12, with
    ↪ B^2 = -1

    # Define a rotation angle (in radians)
    theta = math.pi / 2  # 90 degrees rotation

    # Compute the rotor corresponding to this rotation
    R = compute_rotor(B, theta)
    print("Rotor R =", R)

    # Verify rotor normalization: R * R.reverse() should yield 1
    ↪ (the scalar identity)
```

348

```python
norm_R = R * R.reverse()
print("Rotor normalization (should be 1):", norm_R)

# Define a vector v (pure grade-1 multivector) to be rotated:
# For instance, v = 1e1 + 2e2
v = GA2D(0, 1, 2, 0)
print("Original vector v =", v)

# Apply the rotor to rotate the vector using the sandwich
↪  product
v_rotated = apply_rotor(v, R)
print("Rotated vector v' =", v_rotated)

# Demonstrate successive rotations:
# Compute another rotor for the same angle
R2 = compute_rotor(B, theta)
# Combine the two rotors; the combined rotor represents a
↪  180-degree rotation.
R_total = combine_rotors(R, R2)
print("Combined rotor R_total =", R_total)

# Apply the combined rotor to the original vector
v_rotated_total = apply_rotor(v, R_total)
print("Vector after two successive rotations v'' =",
↪  v_rotated_total)
```

Chapter 50

Phase Space in GA: Representation and Simulation

Multivector Structure of Phase Space

Phase space, as traditionally defined in classical mechanics, encompasses the full set of dynamical variables represented by coordinates and their conjugate momenta. Within the framework of geometric algebra, this double structure finds an elegant unification into a single multivector. A typical representation takes the form

$$X = q + I p,$$

where q represents the position vector, p the momentum vector, and I denotes the pseudoscalar element associated with the algebra. The pseudoscalar I squares to either -1 or $+1$, depending on the metric signature, and serves to couple the two distinct physical quantities into one algebraic object. In this representation, the natural geometric product consolidates rotational symmetries and provides a compact description of phase space dynamics. Such a formulation is particularly advantageous for systems where rotations and reflections have nontrivial impacts on the evolution of both position and momentum.

The use of multivectors to represent phase space enables the direct application of the geometric product and other GA operations to propagate the state. Conservation properties, such as phase space volume preservation, are inherently respected due to the algebraic structure, which is encoded in the non-commutative nature of the products between the different grades. The combined representation further allows for transformations that simultaneously mix the position and momentum components in a way that parallels canonical transformations in classical Hamiltonian mechanics.

Computational Strategies for Phase Space Simulation

The simulation of phase space dynamics in the geometric algebra framework involves evolving the multivector state X over time according to the governing differential equations. The dynamics can be encapsulated in a multivector-valued derivative

$$\frac{dX}{dt} = F(X),$$

where F is an operator defined entirely within GA. Numerical integration schemes, such as Euler's method or higher-order Runge-Kutta algorithms, can be adapted to accommodate the multivector structure. The inherent geometric product, with its non-commutative properties, requires that any computational algorithm must carefully preserve the algebraic relations—ensuring that invariants, such as the normalization of certain rotor components or phase space volume, remain intact.

A correctly chosen numerical method must take into account the interplay between the scalar, vector, and higher-grade components of the multivector state. This ensures that the simulation accurately reflects the theoretical dynamics. The computation is further streamlined by algorithms designed to perform the geometric product and other GA operations with high efficiency, a crucial factor in simulating systems with many degrees of freedom or when long-term integrations are required.

1 Python Implementation of a Phase Space Update Function

The following Python function demonstrates a basic Euler integration step for updating the phase space multivector state. This function is designed to compute the state after a small time step

h, given a callable that returns the multivector derivative based on the current state.

```
def update_phase_space(state, h, derivative_func):
    """
    Compute a single Euler integration step for a phase space
    ↪ multivector.

    Parameters:
        state: A multivector representing the current phase space
        ↪ state,
                which encapsulates both position and momentum
                ↪ components.
        h:      The time step for the integration (a scalar value).
        derivative_func: A callable function that computes the time
        ↪ derivative
                        of the state. It accepts the current state
                        ↪ as input
                        and returns a multivector representing
                        ↪ dX/dt.

    Returns:
        A multivector representing the updated phase space state
        ↪ after
            advancing by the time step h.
    """
    return state + h * derivative_func(state)
```

This function implements a straightforward Euler integration method. It computes the derivative of the state using the supplied function and updates the state by scaling this derivative by the time step h. Although Euler's method is simple, the inclusion of a multivector derivative within the geometric algebra framework demands careful handling of the algebraic operations. The function is intended to serve as a building block upon which more sophisticated numerical schemes can be constructed, all while respecting the intricate structure of phase space as encoded in geometric algebra.

Python Code Snippet

```
# Import required libraries
import numpy as np
import matplotlib.pyplot as plt

# ======================================================
```

```python
# Define a Multivector class to represent the phase space state in
↪   GA.
# In this simplified setup, we use a reduced geometric algebra where
↪   a multivector
# is expressed as:
#     X = q + I * p
# where:
#     q is the scalar (position) component,
#     p is the pseudoscalar (momentum) component, and
#     I is the pseudoscalar satisfying I**2 = -1.
# This algebra is isomorphic to the complex numbers.
# =======================================================

class Multivector:
    def __init__(self, scalar=0.0, pseudoscalar=0.0):
        # scalar: represents the position (q)
        # pseudoscalar: represents the momentum (p)
        self.scalar = scalar
        self.pseudoscalar = pseudoscalar

    def __add__(self, other):
        if isinstance(other, Multivector):
            return Multivector(self.scalar + other.scalar,
                               self.pseudoscalar +
                               ↪  other.pseudoscalar)
        else:
            raise TypeError("Addition is supported only between
            ↪  Multivector types.")

    def __sub__(self, other):
        if isinstance(other, Multivector):
            return Multivector(self.scalar - other.scalar,
                               self.pseudoscalar -
                               ↪  other.pseudoscalar)
        else:
            raise TypeError("Subtraction is supported only between
            ↪  Multivector types.")

    def __mul__(self, other):
        # Supports multiplication with scalars or another
        ↪  Multivector.
        if isinstance(other, (int, float)):
            return Multivector(self.scalar * other,
                               self.pseudoscalar * other)
        elif isinstance(other, Multivector):
            # Geometric product: (q1 + I p1) * (q2 + I p2) =
            # (q1*q2 - p1*p2) + I * (q1*p2 + p1*q2)
            a, b = self.scalar, self.pseudoscalar
            c, d = other.scalar, other.pseudoscalar
            new_scalar = a*c - b*d    # using I**2 = -1
            new_pseudoscalar = a*d + b*c
            return Multivector(new_scalar, new_pseudoscalar)
        else:
```

```python
        raise TypeError("Unsupported multiplication operand
        ↪   type.")

    def __rmul__(self, other):
        # Enable multiplication from left with a scalar.
        return self.__mul__(other)

    def __repr__(self):
        # String representation in the form "q + pI"
        return f"{self.scalar} + {self.pseudoscalar}I"


# ======================================================
# Define the dynamical derivative function F(X) for phase space.
# For a simple harmonic oscillator, the standard equations are:
#     dq/dt = p
#     dp/dt = -q
# In GA form:
#     dX/dt = p - I * q
# where X = q + I * p.
# ======================================================

def derivative_func(state):
    q = state.scalar
    p = state.pseudoscalar
    # Return the derivative multivector: F(X) = p - I * q
    return Multivector(p, -q)


# ======================================================
# Euler integration update function for advancing the phase space
# ↪   state.
# The update rule is:
#     X(t + h) = X(t) + h * F(X(t))
# This method respects the multivector structure in the GA
# ↪   framework.
# ======================================================

def update_phase_space(state, h, derivative_func):
    """
    Compute a single Euler integration step for a phase space
    ↪   multivector.

    Parameters:
        state: A Multivector representing the current phase space
        ↪   state.
        h: The time step for integration.
        derivative_func: A callable that computes dX/dt from the
        ↪   current state.

    Returns:
        A new Multivector representing the updated phase space
        ↪   state.
```

```
    """
    return state + h * derivative_func(state)


# ========================================================
# Main simulation of phase space dynamics (Harmonic Oscillator)
# ========================================================

# Simulation parameters
h = 0.01                # Time step
num_steps = 1000        # Total number of integration steps
time_array = np.linspace(0, h*num_steps, num_steps+1)

# Initial phase space state: for example, start with q(0)=1.0 and
↪  p(0)=0.0.
initial_state = Multivector(1.0, 0.0)

# Store the history of the state variables for plotting
q_values = [initial_state.scalar]
p_values = [initial_state.pseudoscalar]

# Initialize the current state
current_state = initial_state

# Time integration loop using Euler's method
for _ in range(num_steps):
    current_state = update_phase_space(current_state, h,
    ↪  derivative_func)
    q_values.append(current_state.scalar)
    p_values.append(current_state.pseudoscalar)


# ========================================================
# Plotting the results: Phase space trajectory and time evolution.
# ========================================================

# Plot the phase space trajectory (q vs. p)
plt.figure(figsize=(8, 6))
plt.plot(q_values, p_values, label='Phase Space Trajectory')
plt.xlabel('Position q')
plt.ylabel('Momentum p')
plt.title('Phase Space Trajectory of a Harmonic Oscillator')
plt.legend()
plt.grid(True)
plt.show()

# Plot the time evolution of position and momentum separately
plt.figure(figsize=(8, 6))

plt.subplot(2, 1, 1)
plt.plot(time_array, q_values, 'b-', label='Position q')
plt.xlabel('Time')
plt.ylabel('q')
plt.title('Time Evolution of Position')
```

```python
plt.legend()
plt.grid(True)

plt.subplot(2, 1, 2)
plt.plot(time_array, p_values, 'r-', label='Momentum p')
plt.xlabel('Time')
plt.ylabel('p')
plt.title('Time Evolution of Momentum')
plt.legend()
plt.grid(True)
plt.tight_layout()
plt.show()
```

Chapter 51

Fourier Transforms in GA: Computational Techniques

Fundamentals of Fourier Analysis in Geometric Algebra

Fourier analysis provides a powerful tool for decomposing signals into their constituent frequency components. Within the framework of geometric algebra, the conventional Fourier transform undergoes an extension whereby the imaginary unit is replaced by a pseudoscalar element, I, satisfying $I^2 = -1$. This replacement imbues the transform with an innate capability to handle multivector-valued functions. Given a multivector function $f(\mathbf{x})$, the Fourier transform is expressed as

$$\mathcal{F}\{f\}(\mathbf{k}) = \int_{\mathbb{R}^n} f(\mathbf{x})\, e^{-I\mathbf{k}\cdot\mathbf{x}}\, d^n\mathbf{x},$$

where the exponential kernel $e^{-I\mathbf{k}\cdot\mathbf{x}}$ is defined via its series expansion. This formulation preserves the inherent phase and amplitude characteristics, while simultaneously converting contributions across different grades in the algebra.

Mathematical Formulation and Exponential Representation

The transformation kernel in conventional Fourier analysis, e^{-ikx}, finds its counterpart in geometric algebra as e^{-Ikx}. The adaptation is facilitated by the generalized Euler identity in geometric algebra:

$$e^{-Ikx} = \cos(kx) - I\sin(kx).$$

Here, the pseudoscalar I acts as the complex unit and enables the seamless integration of scalar and bivector components during the transformation process. When applied to a multivector-valued function, the Fourier transform decomposes the function into projections along the frequency axis, while preserving the geometric product structure. This method directly exploits the orthogonality properties of the exponential kernels and ensures the invariance of key features in the transformed domain.

Computational Implementation of the Fourier Transform in GA

To transition from the continuous integral representation to a computational algorithm, discretization of the Fourier transform is paramount. The continuous integral,

$$\mathcal{F}\{f\}(k) = \int_{\mathbb{R}} f(x)\, e^{-Ikx}\, dx,$$

is approximated by the discrete summation

$$\mathcal{F}\{f\}(k) \approx \sum_{j=0}^{N-1} f(x_j)\, e^{-Ikx_j}\, \Delta x,$$

where Δx denotes the sampling interval. In this discrete context, the evaluation of the kernel e^{-Ikx_j} is performed through either polynomial series approximations or dedicated routines that compute the geometric exponential.

The following Python function illustrates a computational approach to the Fourier transform for a discretized GA signal. This function processes a list of multivector samples corresponding to the spatial domain and returns their frequency domain representation computed at an array of frequency values.

```python
def compute_fourier_transform(signal, freqs, dx):
    """
    Computes the Fourier transform of a discretized GA signal.

    Parameters:
        signal: A list of multivector objects representing the GA
        ↪ signal.
        freqs: A sequence of frequency values at which the
        ↪ transform is evaluated.
        dx:    The uniform spacing between samples in the spatial
        ↪ domain.

    Returns:
        A list of multivector objects representing the Fourier
        ↪ transform of the input signal.
    """
    transformed = []
    for k in freqs:
        accumulator = Multivector(0.0, 0.0)
        for j, sample in enumerate(signal):
            # Compute the exponential kernel using a GA-specific
            ↪ function.
            # The exponential exp(-I * k * j * dx) is computed based
            ↪ on the series expansion,
            # where I is the pseudoscalar (I**2 = -1).
            kernel = ga_exp(-1j * k * j * dx)  # Placeholder for GA
            ↪ exponential function.
            accumulator = accumulator + sample * kernel
        transformed.append(accumulator * dx)
    return transformed
```

In this implementation, the variable `dx` represents the spatial discretization interval, ensuring that the summation approximates the continuous Fourier integral. The function `ga_exp` is intended as a specialized method for computing the geometric exponential and must account for the intricacies of the geometric product. The accumulator combines the contributions from each sample, and the final multiplication by `dx` completes the numerical integration. This modular approach establishes a framework capable of handling the multivector structure inherent in GA while efficiently transitioning to the frequency domain for analysis of signals and functions.

Python Code Snippet

```python
import math
import numpy as np
```

```python
# Define a simple Multivector class representing scalar and
↪ pseudoscalar parts.
# Here we assume a minimal GA with only the scalar (grade-0) and
↪ pseudoscalar (grade-n) components.
class Multivector:
    def __init__(self, scalar=0.0, pseudoscalar=0.0):
        self.scalar = scalar
        self.pseudoscalar = pseudoscalar

    def __add__(self, other):
        if isinstance(other, Multivector):
            return Multivector(self.scalar + other.scalar,
                               self.pseudoscalar +
                               ↪ other.pseudoscalar)
        else:
            # Allow addition with a scalar (treated as grade-0
            ↪ component)
            return Multivector(self.scalar + other,
            ↪ self.pseudoscalar)

    __radd__ = __add__

    def __sub__(self, other):
        if isinstance(other, Multivector):
            return Multivector(self.scalar - other.scalar,
                               self.pseudoscalar -
                               ↪ other.pseudoscalar)
        else:
            return Multivector(self.scalar - other,
            ↪ self.pseudoscalar)

    def __mul__(self, other):
        # Geometric product for multivectors of the form a + bI,
        ↪ where I^2 = -1.
        # For two multivectors M = a + bI and N = c + dI, we have:
        #     M * N = (ac - bd) + (ad + bc)I.
        if isinstance(other, Multivector):
            new_scalar = self.scalar * other.scalar -
            ↪ self.pseudoscalar * other.pseudoscalar
            new_pseudoscalar = self.scalar * other.pseudoscalar +
            ↪ self.pseudoscalar * other.scalar
            return Multivector(new_scalar, new_pseudoscalar)
        else:
            # Scalar multiplication
            return Multivector(self.scalar * other,
            ↪ self.pseudoscalar * other)

    __rmul__ = __mul__

    def __str__(self):
        return f"{self.scalar} + {self.pseudoscalar}I"

    def __repr__(self):
```

```python
        return self.__str__()

# Global pseudoscalar element I (with I^2 = -1).
I = Multivector(0.0, 1.0)

def ga_exp(M, terms=20):
    """
    Computes the exponential of a multivector M using a series
      expansion.

    For a pure pseudoscalar M = b*I (i.e. with negligible scalar
      part),
    the closed-form result is given by:
        exp(b*I) = cos(b) + I*sin(b)
    Otherwise, the exponential is computed via the power series:
        exp(M) = sum_{n=0}^{} M^n / n!
    """
    # If M is a pure pseudoscalar (scalar nearly zero), use the
    #   closed-form Euler formula.
    if abs(M.scalar) < 1e-12:
        b = M.pseudoscalar
        return Multivector(math.cos(b), math.sin(b))

    # Otherwise, use the series expansion for a general multivector.
    term = Multivector(1.0, 0.0)   # term_0 = 1
    result = Multivector(1.0, 0.0)
    for n in range(1, terms):
        term = term * (M * (1.0 / n))
        result = result + term
    return result

def compute_fourier_transform(signal, freqs, x_values):
    """
    Computes the Fourier transform of a discretized GA signal.

    Parameters:
        signal:   A list of Multivector objects representing the GA
          signal.
        freqs:    A sequence of frequency values at which the
          transform is evaluated.
        x_values: An array of spatial domain sample points
          corresponding to the signal.

    Returns:
        A list of Multivector objects representing the Fourier
          transform of the input signal.
    """
    dx = x_values[1] - x_values[0]
    transformed = []
    for k in freqs:
        accumulator = Multivector(0.0, 0.0)
        for j, sample in enumerate(signal):
            x = x_values[j]
```

```
            # Compute the exponential kernel: exp(-I * k * x)
            # Using the fact that for a pure pseudoscalar argument,
            # exp(-I * k * x) = cos(k * x) - I*sin(k * x).
            kernel = ga_exp(I * (-k * x))
            accumulator = accumulator + sample * kernel
        transformed.append(accumulator * dx)
    return transformed

# Example usage:
if __name__ == "__main__":
    # Spatial domain parameters.
    N = 128
    dx = 0.1
    # Create a spatial grid centered around zero.
    x_values = np.linspace(-N//2 * dx, N//2 * dx, N)

    # Define a sample GA signal: a scalar Gaussian function.
    signal = []
    for x in x_values:
        value = math.exp(-x**2)   # Gaussian profile.
        # Represent the function as a multivector with only a scalar
        ↪       part.
        signal.append(Multivector(value, 0.0))

    # Define a set of frequency values to evaluate the transform.
    freqs = np.linspace(-10, 10, 256)

    # Compute the Fourier transform of the signal.
    ft_signal = compute_fourier_transform(signal, freqs, x_values)

    # Output the first few Fourier transform results.
    for i in range(5):
        print(f"Frequency {freqs[i]:.2f}: {ft_signal[i]}")
```

Chapter 52

Function Spaces and GA: Bridging Analysis and Geometry

Integration of Function Spaces in Geometric Algebra

In many areas of physics, solutions of differential equations and features of quantum systems are represented by functions that reside in infinite-dimensional spaces. Geometric algebra extends the conventional analysis by allowing functions, defined on domains in $\mathbb{R}^n$, to be interpreted as mappings

$$f : \mathbb{R}^n \to \mathcal{G},$$

where $\mathcal{G}$ denotes the geometric algebra associated with the space. In this formulation, each point in the domain is assigned a multivector, whose grade composition encapsulates both magnitude and geometric orientation. The resulting framework is capable of encoding classical function spaces, such as the Hilbert space $L^2(\mathbb{R}^n)$, and augmenting them with operations intrinsic to the geometric product.

The integration of function spaces with GA permits the simultaneous manipulation of scalar, vector, and higher-grade components. In this approach, a function is not merely an element of an abstract vector space, but a rich multivector field whose pointwise operations comply with the rules of GA. This merging of analysis and geometry is particularly advantageous when handling symmetry operations and rotation invariance in physical models.

Operators and Algebraic Structures in Function Spaces

The algebraic operations that define geometric algebra extend naturally into the realm of function spaces. Addition and scalar multiplication are defined pointwise for functions, while the geometric product is applied to the multivector values at each point in the domain. As an example, consider two functions f and g such that

$$f(x), g(x) \in \mathcal{G} \quad \text{for all } x \in \mathbb{R}^n.$$

Their pointwise geometric product is given by

$$(fg)(x) = f(x)g(x),$$

which respects the non-commutative structure and grade selection rules inherent in GA.

Differential operators, such as the gradient or directional derivative, are also recast into the GA language. The derivative of a multivector field

$$\nabla f(x) = \sum_{i=1}^{n} e_i \, \partial_{x_i} f(x),$$

where $\{e_i\}$ forms an orthonormal basis for $\mathbb{R}^n$, acts compatibly with the geometric structure of the function. The operator ∇ simultaneously encodes both divergence and curl when applied to vector-valued functions, thereby providing a unified treatment of many differential operations traditionally treated separately.

An inner product on a GA function space is defined in a way that generalizes the classical L^2 inner product. For functions f and g, an inner product may be written as

$$\langle f, g \rangle = \int_{\mathbb{R}^n} f(x) \widetilde{g(x)} \, d^n x,$$

where $\widetilde{g(x)}$ denotes an appropriate involution (such as reversion) that ensures the product yields a scalar quantity. This definition preserves the geometric interpretation and enables projections onto subspaces defined by multivector basis functions.

Analytic Expansion and Basis Representations

Representations of function spaces benefit from the concept of basis expansion, which in GA takes on an enriched form. A complete set of orthogonal multivector-valued functions $\{\phi_\alpha(x)\}$ can serve as a basis for the space. Every function $f(x)$ then admits an expansion

$$f(x) = \sum_\alpha c_\alpha \, \phi_\alpha(x),$$

with coefficients computed via the inner product

$$c_\alpha = \frac{\langle f, \phi_\alpha \rangle}{\langle \phi_\alpha, \phi_\alpha \rangle}.$$

Such expansions are particularly useful when the basis functions $\phi_\alpha(x)$ are eigenfunctions of differential operators that admit a physical interpretation in models of electromagnetism, quantum mechanics, or fluid dynamics.

This technique provides a framework to analyze the behavior of complex systems by decomposing multivector fields into components associated with distinct geometric features. The traditional separation of amplitude and phase in complex analysis is extended to include contributions from vector and higher-grade components, yielding a more comprehensive method of analysis in physics.

Computational Implementation of GA Function Spaces

When transitioning to computational implementations, discretization of the continuous function space is performed. Sampled values of a GA-valued function, defined on a grid of points in the domain, enable numerical approximations of the inner products and differential operators discussed earlier. A fundamental operation in this context is the computation of the inner product between two discretized functions. The following Python function illustrates a method for performing this calculation in a manner consistent with geometric algebra operations.

```
def compute_ga_inner_product(f_vals, g_vals, dx):
    """
```

```
Compute the inner product of two discretized GA-valued
↪ functions.

The inner product is defined as:
    <f, g> = sum(f_vals[i] * g_vals[i]) * dx
where the multiplication denotes the geometric product of
↪ multivectors.

Parameters:
    f_vals : list of GA multivectors representing samples of the
    ↪ first function.
    g_vals : list of GA multivectors representing samples of the
    ↪ second function.
    dx     : float, the uniform spacing between sample points.

Returns:
    GA multivector representing the inner product.
"""
result = Multivector(0.0, 0.0)
for f, g in zip(f_vals, g_vals):
    result += f * g
return result * dx
```

In this implementation, the discrete inner product approximates the continuous integral by summing the contributions of the geometric product over the sampling domain. The routine assumes the existence of a well-defined Multivector class consistent with the GA framework, and that the geometric product operation is implemented accordingly. Through such computational tools, the analysis of function spaces within the geometric algebraic setting becomes not only conceptually elegant but also practically accessible for numerical analysis in physics.

Python Code Snippet

```
import numpy as np

class Multivector:
    """
    A simple implementation for 2D geometric algebra elements.
    Each multivector is represented as:
        A = a + v1*e1 + v2*e2 + b*(e1e2),
    where 'a' is the scalar part, 'v' is the vector part (a
    ↪ 2-element array),
    and 'b' is the bivector part.
    """
    def __init__(self, scalar=0.0, vector=None, bivector=0.0):
        self.scalar = float(scalar)
```

```python
        if vector is None:
            self.vector = np.zeros(2, dtype=float)
        else:
            self.vector = np.array(vector, dtype=float)
        self.bivector = float(bivector)

    def __add__(self, other):
        if isinstance(other, Multivector):
            return Multivector(self.scalar + other.scalar,
                               self.vector + other.vector,
                               self.bivector + other.bivector)
        else:
            return NotImplemented

    def __radd__(self, other):
        return self.__add__(other)

    def __sub__(self, other):
        if isinstance(other, Multivector):
            return Multivector(self.scalar - other.scalar,
                               self.vector - other.vector,
                               self.bivector - other.bivector)
        else:
            return NotImplemented

    def __mul__(self, other):
        # Scalar multiplication
        if isinstance(other, (int, float)):
            return Multivector(self.scalar * other,
                               self.vector * other,
                               self.bivector * other)
        # Geometric product between two multivectors in 2D GA.
        elif isinstance(other, Multivector):
            # Write self = a + v + b*I and other = c + w + d*I,
            # where I = e1e2 and I^2 = -1.
            a = self.scalar
            b = self.bivector
            v = self.vector
            c = other.scalar
            d = other.bivector
            w = other.vector

            # Scalar part: a*c + dot(v, w) - b*d
            dot_vw = np.dot(v, w)
            new_scalar = a * c + dot_vw - b * d

            # The vector part:
            # v * c + a * w  are straightforward.
            # Additionally, bivector times vector: I*w =
            ↪   np.array([-w[1], w[0]])
            # and vector times bivector: v*I = np.array([v[1],
            ↪   -v[0]])
```

```python
            new_vector = a * w + c * v + b * np.array([-w[1], w[0]])
                + d * np.array([v[1], -v[0]])

            # Bivector part:
            # Comes from a*d + b*c + wedge(v, w)
            wedge = v[0] * w[1] - v[1] * w[0]
            new_bivector = a * d + b * c + wedge

            return Multivector(new_scalar, new_vector, new_bivector)
        else:
            return NotImplemented

    def __rmul__(self, other):
        # Allow scalar multiplication from left
        return self.__mul__(other)

    def __iadd__(self, other):
        temp = self + other
        self.scalar, self.vector, self.bivector = temp.scalar,
            temp.vector, temp.bivector
        return self

    def __repr__(self):
        return (f"Multivector(scalar={self.scalar}, "
            f"vector={self.vector.tolist()}, "
            f"bivector={self.bivector})")

    def reverse(self):
        """
        The reversal operation in 2D GA: the scalar and vector parts
            remain unchanged,
        while the bivector part changes sign.
        """
        return Multivector(self.scalar, self.vector, -self.bivector)

def compute_ga_inner_product(f_vals, g_vals, dx):
    """
    Compute the inner product of two discretized GA-valued
        functions.

    The inner product is defined as:
        <f, g> =  f(x) (g(x))~ dx
    where (g(x))~ is the reversal of g(x), and the geometric product
        is used.
    In a discretized setting, the integral is approximated by a sum:

        <f, g>  sum [ f_vals[i] * (g_vals[i]).reverse() ] * dx
    """
    result = Multivector(0.0, [0.0, 0.0], 0.0)
    for f, g in zip(f_vals, g_vals):
        result += f * g.reverse()
    return result * dx
```

```python
def finite_difference_gradient(f_vals, dx):
    """
    Approximate the gradient (derivative) of a discretized GA-valued
    ↪ function in 1D.
    For each sample point, use central differences where possible
    ↪ and one-sided differences at the boundaries.
    """
    grad_vals = []
    n = len(f_vals)
    for i in range(n):
        if i == 0:
            # Forward difference
            diff = (f_vals[i+1] - f_vals[i]) * (1.0 / dx)
        elif i == n - 1:
            # Backward difference
            diff = (f_vals[i] - f_vals[i-1]) * (1.0 / dx)
        else:
            # Central difference
            diff = (f_vals[i+1] - f_vals[i-1]) * (0.5 / dx)
        grad_vals.append(diff)
    return grad_vals

def compute_basis_coefficients(f_vals, basis_funcs, dx):
    """
    Given a discretized GA-valued function f_vals and a list of
    ↪ basis functions
    (each basis function is given as a list of Multivector values
    ↪ sampled on the grid),
    compute the expansion coefficients using the inner product.

    Each coefficient is computed as:
        c_alpha = <f, phi_alpha> / <phi_alpha, phi_alpha>,
    where only the scalar part of the denominator is used for
    ↪ simplicity.
    """
    coeffs = []
    for phi in basis_funcs:
        inner_fp = compute_ga_inner_product(f_vals, phi, dx)
        inner_phi = compute_ga_inner_product(phi, phi, dx)
        coeff = inner_fp.scalar / inner_phi.scalar if
        ↪ inner_phi.scalar != 0 else 0.0
        coeffs.append(coeff)
    return coeffs

if __name__ == "__main__":
    # Create a 1D grid over the interval [0, 10] with 100 sample
    ↪ points
    x_vals = np.linspace(0, 10, 100)
    dx = x_vals[1] - x_vals[0]

    # Define two GA-valued functions f(x) and g(x)
    # Here, f(x) = 1 + x*e1 + 0.5*sin(x)*(e1e2)
    #       g(x) = 2 + 0.5*x*e1 + 0.3*cos(x)*(e1e2)
```

```python
f_samples = []
g_samples = []
for x in x_vals:
    f_val = Multivector(1.0, [x, 0.0], 0.5 * np.sin(x))
    g_val = Multivector(2.0, [0.5 * x, 0.0], 0.3 * np.cos(x))
    f_samples.append(f_val)
    g_samples.append(g_val)

# Compute and display the inner product of f and g
inner_product = compute_ga_inner_product(f_samples, g_samples,
↪    dx)
print("Inner Product of f and g:", inner_product)

# Compute and display the finite-difference gradient of f
↪    (approximation of derivative)
grad_f = finite_difference_gradient(f_samples, dx)
midpoint_index = len(grad_f) // 2
print("Gradient of f at midpoint:", grad_f[midpoint_index])

# Define example basis functions for expansion:
# Basis 1: constant function, phi_0(x) = 1
# Basis 2: linear function, phi_1(x) = x*e1 (only vector part
↪    along e1)
basis_const = [Multivector(1.0) for _ in x_vals]
basis_linear = [Multivector(0.0, [x, 0.0]) for x in x_vals]
basis_funcs = [basis_const, basis_linear]

# Compute expansion coefficients of f relative to the basis
↪    functions
coeffs = compute_basis_coefficients(f_samples, basis_funcs, dx)
print("Basis Expansion Coefficients for f:", coeffs)
```

Chapter 53

Integration Techniques: GA Methods for Computation

Numerical Integration Methods in Geometric Algebra

Within the framework of geometric algebra, numerical integration is performed over functions that map from the domain in $\mathbb{R}^n$ to the geometric algebra $\mathcal{G}$. The integration of a GA-valued function is implemented by decomposing the multivector into its constituent grade components and applying conventional quadrature rules directly to each of these components. For a discretized function

$$F : \mathbb{R}^n \to \mathcal{G},$$

the standard trapezoidal rule can be extended to approximate the integral as

$$\int_a^b F(x)\,dx \approx \frac{dx}{2}\left(F(a) + 2\sum_{i=1}^{N-1} F(x_i) + F(b)\right),$$

where dx is the step length and the summation is performed in accordance with the linearity of integration. This approach ensures that each component of the multivector is integrated while retaining the noncommutative structure inherent in the geometric product.

A representative implementation of the trapezoidal integration rule for GA-valued functions is provided by the following Python function:

```
def ga_trapezoidal_integration(f_vals, dx):
    """
    Approximate the integral of a discretized GA-valued function
    ↪ using the trapezoidal rule.

    Parameters:
        f_vals : list of GA multivectors representing function
        ↪ samples.
        dx     : float, the spacing between successive sample
        ↪ points.

    Returns:
        GA multivector approximately equal to the definite integral.
    """
    if not f_vals:
        return None
    integral = 0.5 * (f_vals[0] + f_vals[-1])
    for f in f_vals[1:-1]:
        integral += f
    return integral * dx
```

The function ga_trapezoidal_integration demonstrates the basic principle of accumulating the contributions from each sample point while applying appropriate weighting to the endpoints. The integration process respects the algebraic structure by relying on the addition and scalar multiplication rules defined in the GA framework.

Symbolic Integration Techniques in GA Function Spaces

Symbolic integration in geometric algebra involves the decomposition of a multivector function into its basis components and the subsequent application of standard rules from symbolic calculus. If a multivector function is expressed in the form

$$F(x) = \sum_i f_i(x) B_i,$$

with each B_i representing a basis blade and $f_i(x)$ a scalar function, then the integral of $F(x)$ with respect to x is obtained by evaluating

$$\int F(x)\,dx = \sum_i \left(\int f_i(x)\,dx\right) B_i.$$

This formulation leverages the linearity of integration and treats the basis blades as constants. The facility to manage and manipulate each scalar function $f_i(x)$ symbolically permits the derivation of closed-form integrals or the application of series expansion techniques when elementary antiderivatives are not available.

The symbolic approach is particularly effective when the functional dependencies are analytically tractable. The techniques from computer algebra systems, equipped to handle polynomial, transcendental, or rational functions, can be adapted to integrate multivector fields by considering each grade independently. Such algorithms must properly account for the geometric product rules during the integration process to ensure that the resulting expressions maintain the geometric meaning.

Algorithmic Frameworks for GA Integration

The computational implementation of integration techniques within geometric algebra necessitates careful algorithmic design. Efficient integration algorithms must manage the inherent noncommutativity and multigrade structure of GA while optimizing computational resources. Data structures representing multivectors are constructed to support rapid arithmetic operations, including addition, scalar multiplication, and the geometric product.

Adaptive quadrature methods can be utilized to improve accuracy while minimizing computational cost. The algorithmic framework typically involves precomputing weights and nodes for integration schemes, performing the integration on each grade of the multivector, and recombining the results to form the integrated multivector. The design of these algorithms hinges upon error analysis and convergence properties of the numerical schemes, particularly in applications where the integration is performed over a high-dimensional domain.

The integration of differential operators in geometric calculus further benefits from the synergy between symbolic manipulation and numerical approximation. The use of specialized codes that encapsulate the algebraic rules of GA allows for the automation of integration routines. These codes must ensure that the structural properties of the geometric algebra, such as grade involution and reversion, are preserved during the integration. The blend of numerical and symbolic methods thus provides a comprehensive toolkit for addressing complex integrations in computational physics.

Python Code Snippet

```
import math
import sympy as sp

# Helper function: Convert a basis blade string to a tuple of
↪ integers.
def basis_to_tuple(basis):
    """
    Given a basis blade string, returns a tuple of integers
    ↪ representing the blade.
    For example, "1" -> (), "e1" -> (1,), "e12" -> (1,2).
    Assumes that indices are single-digit and in increasing order.
    """
    if basis == "1":
        return ()
    # Remove the leading 'e' and convert each character to an
    ↪ integer.
    return tuple(int(ch) for ch in basis[1:])

# Helper function: Multiply two basis blades.
def multiply_basis(b1, b2):
    """
    Computes the product of two basis blades b1 and b2.
    The rule is based on the geometric product for a Euclidean
    ↪ metric where e_i^2 = 1.
    It returns a tuple (result_basis, sign) where:
      - result_basis is a string representing the resulting basis
      ↪ (e.g. "e12")
      - sign is the factor from reordering basis elements.
    """
    a = basis_to_tuple(b1)
    b = basis_to_tuple(b2)
    L = list(a)
    sign = 1
    # Process each element in b.
    for bi in b:
        if bi in L:
            # Remove the duplicate; e_i^2 = 1 so the pair cancels.
            L.remove(bi)
```

```python
        else:
            # Count the number of elements greater than bi in L to
            ↪   determine inversions.
            inversions = sum(1 for x in L if x > bi)
            sign *= (-1) ** inversions
            L.append(bi)
            L.sort()  # Ensure the blade is in sorted order.
    # Convert the tuple back to a string representation.
    if not L:
        result = "1"
    else:
        result = "e" + "".join(str(x) for x in L)
    return result, sign

# Class representing a geometric algebra multivector.
class GAMultivector:
    def __init__(self, components):
        """
        Initialize a GAMultivector.
        components: dict mapping basis blades (as strings, e.g.,
        ↪   "1", "e1", "e12") to scalars
                    (could be numerical or sympy expressions).
        """
        self.components = components.copy()

    def __add__(self, other):
        new_components = self.components.copy()
        for k, v in other.components.items():
            new_components[k] = new_components.get(k, 0) + v
        return GAMultivector(new_components)

    def __radd__(self, other):
        # This supports sum() where the start value may be 0.
        if other == 0:
            return self
        return self.__add__(other)

    def __mul__(self, other):
        if isinstance(other, (int, float, sp.Expr)):
            # Scalar multiplication.
            return GAMultivector({k: v * other for k, v in
            ↪   self.components.items()})
        elif isinstance(other, GAMultivector):
            # Geometric product of two multivectors.
            result = {}
            for b1, val1 in self.components.items():
                for b2, val2 in other.components.items():
                    new_basis, sign = multiply_basis(b1, b2)
                    result[new_basis] = result.get(new_basis, 0) +
                    ↪   val1 * val2 * sign
            return GAMultivector(result)
        else:
            raise TypeError("Unsupported multiplication type.")
```

```python
    def __rmul__(self, other):
        return self.__mul__(other)

    def __str__(self):
        terms = []
        for k, v in self.components.items():
            if v != 0:
                terms.append(f"{v}*{k}")
        return " + ".join(terms) if terms else "0"

    def __repr__(self):
        return self.__str__()

# Numerical Integration using the trapezoidal rule for GA-valued
#   functions.
def ga_trapezoidal_integration(f_vals, dx):
    """
    Approximate the integral of a discretized GA-valued function
      using the trapezoidal rule.

    Parameters:
        f_vals : list of GAMultivector objects representing function
            samples.
        dx     : float, the spacing between successive sample
            points.

    Returns:
        GAMultivector approximately equal to the definite integral.

    The formula applied is:
      _a^b F(x) dx   (dx/2) * (F(a) + 2* F(x_i) + F(b))
    """
    if not f_vals:
        return None
    # Initialize the integral with half of the first and last
    #   samples.
    integral = 0.5 * (f_vals[0] + f_vals[-1])
    # Sum the intermediate samples.
    for sample in f_vals[1:-1]:
        integral = integral + sample
    return integral * dx

# Symbolic integration of a GA-valued function.
def symbolic_integrate_ga(F, x):
    """
    Performs symbolic integration of a GA-valued function F(x) over
      the variable x.

    Parameters:
        F : dict mapping basis blade (string) to sympy expressions
            f(x).
        x : sympy.Symbol representing the integration variable.
```

```
    Returns:
        A dict mapping each basis blade to its integrated sympy
        ↪ expression.

    The integration is performed component-wise:
        F(x) dx =  ( f_i(x) dx) * B_i
    """
    integrated = {}
    for basis, expr in F.items():
        integrated[basis] = sp.integrate(expr, x)
    return integrated

# Function to generate numerical samples of a GA-valued function.
def sample_ga_function(f_scalar_funcs, x_vals):
    """
    Generates a list of GAMultivector samples for the GA-valued
    ↪ function:
        F(x) = ( f_scalar_funcs[basis](x) * (basis) )

    Parameters:
        f_scalar_funcs : dict mapping basis blade (e.g., "1", "e1")
        ↪ to a Python function f(x).
        x_vals         : list of sample points in the domain.

    Returns:
        List of GAMultivector samples evaluated at each x in x_vals.
    """
    samples = []
    for x in x_vals:
        comp = {basis: func(x) for basis, func in
        ↪ f_scalar_funcs.items()}
        samples.append(GAMultivector(comp))
    return samples

# Main demonstration of numerical and symbolic integration in GA.
if __name__ == "__main__":
    # Numerical Integration Demonstration
    a = 0.0
    b = math.pi
    N = 100   # Number of subintervals.
    dx = (b - a) / N
    x_vals = [a + i * dx for i in range(N + 1)]

    # Define a GA-valued function F(x) = x^2 * 1 + sin(x) * e1.
    f_scalar_funcs = {
        "1": lambda x: x**2,
        "e1": lambda x: math.sin(x)
    }

    # Generate numerical samples of F(x).
    samples = sample_ga_function(f_scalar_funcs, x_vals)
```

```python
# Compute the integral using the trapezoidal rule.
numerical_integral = ga_trapezoidal_integration(samples, dx)
print("Numerical Integration Result:")
print(numerical_integral)

# Symbolic Integration Demonstration using sympy.
x = sp.symbols('x')
F_symbolic = {
    "1": x**2,
    "e1": sp.sin(x)
}

symbolic_integral = symbolic_integrate_ga(F_symbolic, x)
print("\nSymbolic Integration Result:")
for basis, expr in symbolic_integral.items():
    print(f"Integral for basis {basis}: {expr}")
```

Chapter 54

Geometric Calculus: Differentiation within GA

Foundations of Differentiation in Geometric Algebra

In geometric calculus the concept of differentiation is extended from real-valued functions to multivector fields defined within a geometric algebra framework. A multivector field is expressed as

$$F(x) = \sum_i f_i(x)\, B_i,$$

where each $f_i(x)$ is a scalar function and each B_i is a basis blade of the algebra. Differentiation in this context is performed by applying the derivative operator to each scalar coefficient independently and then reassembling the result as a multivector. For example, the derivative of $F(x)$ with respect to a variable x is given by

$$\frac{dF}{dx} = \sum_i \frac{df_i(x)}{dx}\, B_i.$$

This componentwise differentiation adheres to the linear structure of the multivector space and guarantees that the resulting derivative preserves the algebraic relationships of the original field. In higher dimensions, the generalization takes the form of a vector derivative,

$$\nabla F(x) = \sum_j e_j \frac{\partial F(x)}{\partial x_j},$$

where the set $\{e_j\}$ represents an orthonormal basis of the underlying Euclidean space. The geometric derivative encapsulates both divergence-like and curl-like operations, unifying various differential concepts under a single operator.

Symbolic Differentiation Techniques in Geometric Algebra

Symbolic differentiation in geometric algebra hinges on the systematic treatment of each scalar component within a multivector field. Computer algebra systems can be employed to perform analytic differentiation on each $f_i(x)$ and, by linearity, the overall derivative is assembled by recombining the results with their corresponding basis blades. A typical implementation involves representing the multivector function as a dictionary mapping each basis blade to a sympy expression.

A fully explained Python function that computes the symbolic derivative componentwise is provided below:

```
def symbolic_differentiate_ga(F, x):
    """
    Compute the symbolic derivative of a GA-valued function F with
    ↪ respect to x.

    Parameters:
        F : dict mapping a basis blade (e.g., "1", "e1") to a sympy
        ↪ expression.
        x : sympy.Symbol with respect to which differentiation is
        ↪ performed.

    Returns:
        A dict representing the derivative, mapping each basis blade
        ↪ to its derivative.
    """
    from sympy import diff
    return {blade: diff(expr, x) for blade, expr in F.items()}
```

This function receives a representation of a GA-valued function $F(x)$ in which each key corresponds to a basis blade and each value is a scalar function. By applying sympy's differentiation routine to each component, the function returns the derivative $\frac{dF}{dx}$ such that

$$\frac{dF}{dx} = \sum_{\text{blade}} \frac{d\left(f_{\text{blade}}(x)\right)}{dx} B_{\text{blade}}.$$

This componentwise approach ensures that the structural integrity of the multivector is maintained while leveraging the power of symbolic computation.

Numerical Differentiation Approaches in Geometric Algebra

In numerous applications in computational physics the analytic form of a derivative may be either unavailable or too complex for closed-form analysis, necessitating numerical differentiation. When a multivector field $F(x)$ is sampled at a discrete set of points, its derivative can be approximated using finite difference techniques. For instance, when the function is defined on a grid with uniform spacing Δx, an interior derivative can be approximated by the central difference formula

$$\frac{dF}{dx}(x_i) \approx \frac{F(x_{i+1}) - F(x_{i-1})}{2\,\Delta x}.$$

For a GA-valued function, each scalar coefficient within the field is treated independently; the derivative of the entire field is obtained by computing the finite difference for each component and then recombining the results with the corresponding basis blades:

$$\frac{dF}{dx}(x_i) \approx \sum_i \frac{f_i(x_{i+1}) - f_i(x_{i-1})}{2\,\Delta x} B_i.$$

This method leverages the linearity of differentiation and is particularly effective when the field is sampled at sufficiently small intervals to ensure convergence of the numerical scheme. Adaptive methods and higher-order finite difference schemes can be integrated into the computational framework to enhance the accuracy of derivative approximations, especially in the presence of steep gradients or nonuniform sampling.

The numerical approach outlined here is essential for simulating dynamics governed by multivector fields in complex physical systems where direct symbolic differentiation is impractical. By discretizing the domain and applying finite difference approximations componentwise, it is possible to achieve an accurate and computationally efficient estimation of the geometric derivative.

Python Code Snippet

```python
import sympy as sp
import numpy as np

def symbolic_differentiate_ga(F, x):
    """
    Compute the symbolic derivative of a GA-valued function F with
       respect to x.

    Parameters:
        F : dict
            Dictionary mapping each basis blade (as a string, e.g.,
               "1", "e1") to a sympy expression.
        x : sympy.Symbol
            The variable with respect to which differentiation is
               performed.

    Returns:
        dict
            A new dictionary mapping each basis blade to its
               corresponding derivative expression.
            That is, dF/dx = sum { diff(f_i(x), x) * B_i }.
    """
    return {blade: sp.diff(expr, x) for blade, expr in F.items()}

def numerical_derivative(arr, dx):
    """
    Compute the numerical derivative of a one-dimensional numpy
       array using finite differences.

    Uses a central difference scheme for interior points and
       forward/backward differences
    for the endpoints.

    Parameters:
        arr : numpy.ndarray
            Array of function values sampled on a uniform grid.
        dx : float
            Spacing between consecutive grid points.

    Returns:
        numpy.ndarray
            Array containing the derivative approximations at each
               grid point.
    """
    d_arr = np.zeros_like(arr)
```

```python
    # Central difference for interior points
    d_arr[1:-1] = (arr[2:] - arr[:-2]) / (2 * dx)
    # Forward difference for the first endpoint
    d_arr[0] = (arr[1] - arr[0]) / dx
    # Backward difference for the last endpoint
    d_arr[-1] = (arr[-1] - arr[-2]) / dx
    return d_arr

def numerical_differentiate_ga(F_values, dx):
    """
    Compute the numerical derivative of a GA-valued function given
    ↪ by F_values.

    The GA-valued function is represented as a dictionary where each
    ↪ key is a basis blade
    and each value is a numpy array containing the evaluations of
    ↪ the corresponding scalar
    component over a grid.

    Parameters:
        F_values : dict
            Dictionary where keys are basis blades (e.g., "1", "e1")
            ↪ and values are numpy arrays.
        dx : float
            Grid spacing used for computing finite differences.

    Returns:
        dict
            New dictionary with the same keys representing the
            ↪ numerical derivative of each component.
    """
    return {blade: numerical_derivative(values, dx) for blade,
    ↪ values in F_values.items()}

if __name__ == "__main__":
    # ---------------------------
    # Symbolic Differentiation
    # ---------------------------

    # Define the symbolic variable.
    x = sp.symbols('x')

    # Define a GA-valued function F(x) = f1(x)*1 + f2(x)*e1,
    # where f1(x) = sin(x)^2 and f2(x) = cos(x).
    F_sym = {
        "1": sp.sin(x)**2,
        "e1": sp.cos(x)
    }

    # Compute the symbolic derivative dF/dx for each component.
    F_sym_deriv = symbolic_differentiate_ga(F_sym, x)

    print("Symbolic Derivative:")
```

```python
for blade, expr in F_sym_deriv.items():
    print(f"Blade {blade}:")
    sp.pprint(expr)
    print()  # newline for clarity

# ---------------------------
# Numerical Differentiation
# ---------------------------

# Create a numerical grid for x in the range [0, 2*pi].
num_points = 100
x_vals = np.linspace(0, 2 * np.pi, num_points)
dx = x_vals[1] - x_vals[0]

# Convert the symbolic expressions in F_sym to numerical
#   functions using lambdify.
f1_lambda = sp.lambdify(x, F_sym["1"], "numpy")
f2_lambda = sp.lambdify(x, F_sym["e1"], "numpy")

# Evaluate the GA-valued function on the grid.
F_num = {
    "1": f1_lambda(x_vals),
    "e1": f2_lambda(x_vals)
}

# Numerically differentiate F_num using finite difference
#   approximations.
F_num_deriv = numerical_differentiate_ga(F_num, dx)

print("Numerical Derivative (Finite Difference Approximation):")
for blade, deriv in F_num_deriv.items():
    print(f"Blade {blade}:")
    print(deriv)
    print()  # newline for clarity
```

Chapter 55

The Laplacian in GA: Computation and Applications

Definition of the Laplacian in Geometric Algebra

In the geometric algebra framework the Laplacian operator is formulated through the repeated application of the vector derivative. Let a multivector field be denoted by

$$F(x) = \sum_i f_i(x) B_i,$$

where each $f_i(x)$ is a scalar function and each B_i is a basis blade. The vector derivative is defined as

$$\nabla = \sum_j e_j \, \partial_{x_j},$$

with $\{e_j\}$ representing an orthonormal basis of the underlying space. The Laplacian is obtained by applying the derivative operator twice, forming the geometric square

$$\nabla^2 = \nabla \cdot \nabla + \nabla \wedge \nabla.$$

Due to the antisymmetric property of the outer product, the bivector component $\nabla \wedge \nabla$ vanishes for functions with continuous second derivatives. Consequently, the Laplacian in this context reduces to

$$\nabla^2 = \nabla \cdot \nabla.$$

For a scalar function $f(x)$ this definition coincides with the conventional notion of the Laplacian,

$$\Delta f = \sum_j \frac{\partial^2 f}{\partial x_j^2},$$

while for a general multivector field the Laplacian is evaluated by applying the operator componentwise and subsequently recombining the results with the corresponding basis blades.

Computational Implementation of the Laplacian

The computational evaluation of the Laplacian operator in geometric algebra necessitates a discretized representation of the derivative operator. Finite difference schemes offer a reliable method for approximating the second derivative in numerical simulations. In a one-dimensional setting with uniform grid spacing Δx, the second derivative of a scalar function $f(x)$ at a grid point x_i is approximated by the central difference formula:

$$\left. \frac{d^2 f}{dx^2} \right|_{x_i} \approx \frac{f(x_{i+1}) - 2f(x_i) + f(x_{i-1})}{\Delta x^2}.$$

When applied within the framework of geometric algebra the finite difference operator is implemented on each scalar component $f_i(x)$ of the multivector field. The recombination with the basis blades then yields an approximation of the Laplacian of the full field.

A Python function developed to compute the numerical Laplacian of a one-dimensional array, using central differences, encapsulates this algorithmic approach. This function demonstrates the finite difference method applied to a discretely sampled scalar function.

```
def numerical_laplacian(arr, dx):
    """
    Compute the numerical Laplacian (second derivative) of a 1-D
    ↪ numpy array using central differences.
```

```
Parameters:
    arr : numpy.ndarray
        Array of scalar function values sampled on a uniform
        ↪ grid.
    dx : float
        Grid spacing between consecutive points.

Returns:
    numpy.ndarray
        Array containing the Laplacian approximation at each
        ↪ grid point.
"""
import numpy as np
lap = np.zeros_like(arr)
# Central difference for interior points
lap[1:-1] = (arr[2:] - 2 * arr[1:-1] + arr[:-2]) / (dx ** 2)
# Forward difference for the first point
lap[0] = (arr[1] - 2 * arr[0] + arr[0]) / (dx ** 2)
# Backward difference for the last point
lap[-1] = (arr[-1] - 2 * arr[-1] + arr[-2]) / (dx ** 2)
return lap
```

By applying the numerical Laplacian function to each scalar component of a multivector field, it is possible to construct a computational procedure that respects the algebraic structure inherent in geometric algebra. This approach allows for the simulation of dynamic processes governed by partial differential equations in a manner that exploits the unified language of GA.

Applications in Physical Systems

The Laplacian operator, as formulated within geometric algebra, finds widespread application in various domains of physics. In quantum mechanics the Laplacian appears in the formulation of the kinetic energy operator of the Schrödinger equation, where it is essential for characterizing the evolution of wave functions. The ability to extend the Laplacian to multivector fields enables the incorporation of additional degrees of freedom, such as spin, within the quantum framework.

In classical field theories the Laplacian is instrumental in describing phenomena such as diffusion, wave propagation, and potential theory. For example, in electrostatics the relation

$$\nabla^2 \phi = -\frac{\rho}{\epsilon_0}$$

expresses the connection between the electric potential ϕ and the charge density ρ, with ϵ_0 denoting the permittivity of free space. Within the geometric algebra framework this equation is naturally extended to accommodate more complex geometrical configurations while maintaining covariance under coordinate transformations.

The computational methods detailed above allow for the robust simulation of physical systems wherein the Laplacian plays a central role. By discretizing the domain and employing finite difference approximations on each scalar component of a GA-valued field, simulations achieve both high accuracy and the preservation of geometric structure. This unified approach streamlines the analysis of diffusion processes, electromagnetic fields, and quantum mechanical systems within a single algebraic framework.

Python Code Snippet

```
import numpy as np
import matplotlib.pyplot as plt

def central_difference(arr, dx):
    """
    Compute the numerical Laplacian (second derivative) of a 1D
     array using the
    central difference scheme. This approximates the second
     derivative as:

    d²f/dx²  (f[i+1] - 2*f[i] + f[i-1]) / dx²

    Parameters:
        arr : numpy.ndarray
            Array of scalar function values sampled on a uniform
             grid.
        dx : float
            The grid spacing between consecutive points.

    Returns:
        numpy.ndarray
            Array containing the approximated second derivative at
             each grid point.
    """
    lap = np.zeros_like(arr)
    # Central difference for interior points:
    lap[1:-1] = (arr[2:] - 2 * arr[1:-1] + arr[:-2]) / (dx ** 2)
    # For the endpoints, use one-sided differences (here,
     forward/backward differences)
    lap[0] = (arr[1] - 2 * arr[0] + arr[0]) / (dx ** 2)
```

```python
        lap[-1] = (arr[-1] - 2 * arr[-1] + arr[-2]) / (dx ** 2)
        return lap

class MultiVectorField:
    """
    Represents a multivector field in the geometric algebra (GA)
    ↪ framework.

    In GA, a multivector field F(x) is decomposed into scalar
    ↪ component functions
    multiplied by basis blades:

        F(x) = _i f_i(x) * B_i

    This class stores each component in a dictionary, where keys are
    ↪ identifiers
    for the basis blades (e.g., 'scalar', 'e1', etc.) and values are
    ↪ the corresponding
    scalar fields (numpy arrays). The Laplacian (²) is applied
    ↪ componentwise.
    """

    def __init__(self, grid, components):
        """
        Initialize the multivector field.

        Parameters:
            grid : numpy.ndarray
                1D array of spatial points.
            components : dict
                Dictionary with keys as basis blade labels and
                ↪ values as 1D numpy arrays
                representing the scalar function for that component.
        """
        self.grid = grid
        self.components = components

    def apply_numerical_laplacian(self, dx):
        """
        Apply the numerical Laplacian to each scalar component of
        ↪ the multivector field.

        Using the finite difference scheme, this method approximates
        ↪ the Laplacian
        operator ² = · on each individual scalar function f_i(x) and
        ↪ returns a
        dictionary of Laplacians for each basis blade.

        Parameters:
            dx : float
                Grid spacing between consecutive points.

        Returns:
```

```
            dict
                Dictionary with the same keys as 'components' and
                ↪   values as the Laplacian
                approximation for each component.
        """
        laplacian_components = {}
        for blade, field in self.components.items():
            laplacian_components[blade] = central_difference(field,
                ↪   dx)
        return laplacian_components

def example_usage():
    """
    Demonstrate the computation of the Laplacian in the framework of
    ↪   geometric algebra.

    The example proceeds as follows:
        1. Define a spatial grid.
        2. Create example scalar functions representing components
        ↪   of a multivector field.
            For instance, a scalar field and a vector field (the e1
            ↪   component) are defined.
            This corresponds to meeting the GA formulation:
                F(x) = f_scalar(x) * 1 + f_e1(x) * e1.
        3. Compute the numerical Laplacian of each component using
        ↪   finite differences.
        4. Plot the original functions and their Laplacians to
        ↪   illustrate the results.
    """
    # Define spatial grid parameters
    x_start, x_end, num_points = 0, 10, 200
    grid = np.linspace(x_start, x_end, num_points)
    dx = grid[1] - grid[0]

    # Example scalar function: f(x) = sin(x)
    f_scalar = np.sin(grid)

    # Another example for a vector component: for instance, f_e1(x)
    ↪   = sin(x)*cos(x)
    f_e1 = np.sin(grid) * np.cos(grid)

    # Dictionary to hold different components of the multivector
    ↪   field.
    # 'scalar' represents the grade-0 component,
    # 'e1' represents a grade-1 component aligned with the basis
    ↪   vector e1.
    components = {
        'scalar': f_scalar,
        'e1': f_e1
    }

    # Instantiate the multivector field
    mv_field = MultiVectorField(grid, components)
```

```python
    # Compute the Laplacian for each component of the multivector
    ↪ field
    laplacians = mv_field.apply_numerical_laplacian(dx)

    # Retrieve Laplacians
    lap_scalar = laplacians['scalar']
    lap_e1 = laplacians['e1']

    # Set up plots to compare the original functions and their
    ↪ Laplacians
    plt.figure(figsize=(12, 8))

    # Plot for scalar component and its Laplacian
    plt.subplot(2, 1, 1)
    plt.plot(grid, f_scalar, label='f_scalar(x) = sin(x)',
    ↪ color='blue')
    plt.plot(grid, lap_scalar, label='Laplacian of f_scalar',
    ↪ linestyle='--', color='red')
    plt.xlabel('x')
    plt.ylabel('Value')
    plt.title('Scalar Component and its Numerical Laplacian')
    plt.legend()
    plt.grid(True)

    # Plot for e1 component and its Laplacian
    plt.subplot(2, 1, 2)
    plt.plot(grid, f_e1, label='f_e1(x) = sin(x)*cos(x)',
    ↪ color='green')
    plt.plot(grid, lap_e1, label='Laplacian of f_e1',
    ↪ linestyle='--', color='magenta')
    plt.xlabel('x')
    plt.ylabel('Value')
    plt.title('Vector Component (e1) and its Numerical Laplacian')
    plt.legend()
    plt.grid(True)

    plt.tight_layout()
    plt.show()

if __name__ == "__main__":
    example_usage()
```

Chapter 56

Electromagnetic Fields in GA: Modeling and Analysis

Mathematical Formalism of Electromagnetic Field Representation

Electromagnetic phenomena can be elegantly encapsulated in geometric algebra by representing the field as a multivector composed of vector and bivector components. In three-dimensional space the electric field is modeled as a vector with components E_1, E_2, and E_3, while the magnetic field is naturally encoded as a bivector. The bivector is constructed from the standard basis vectors through the relation

$$I = e_1 e_2 e_3,$$

so that the electromagnetic field multivector takes the compact form

$$F = E_1 e_1 + E_2 e_2 + E_3 e_3 + B_1(e_2 \wedge e_3) + B_2(e_3 \wedge e_1) + B_3(e_1 \wedge e_2).$$

This formulation is equivalent to the common expression

$$F = \vec{E} + I\vec{B},$$

where the bivector part $I\vec{B}$ represents the magnetic field. The unification provided by geometric algebra allows Maxwell's equations to be written as a single compact equation,

$$\nabla F = J,$$

with $\nabla = e_1 \partial_{x_1} + e_2 \partial_{x_2} + e_3 \partial_{x_3}$ denoting the vector derivative and J representing the four-current. Grade projection operators facilitate the extraction of physical observables from F, yielding the customary divergence and curl operations from the geometric product.

Computational Representation of Electromagnetic Field Bivectors

The efficient computational handling of electromagnetic fields within geometric algebra demands a representation that explicitly preserves the multigrade structure. Digital implementations often employ data structures such as dictionaries or objects that store each grade component of the multivector independently. For example, the electric field vector is directly associated with the grade-1 components, while the magnetic field is embedded as bivector terms. The correspondence is made precise by defining the electromagnetic field as

$$F = E_1 e_1 + E_2 e_2 + E_3 e_3 + B_1(e_2 \wedge e_3) + B_2(e_3 \wedge e_1) + B_3(e_1 \wedge e_2).$$

This structure facilitates the direct implementation of numerical algorithms that respect the underlying geometric relationships. A representative function that constructs the electromagnetic field from independent electric and magnetic component dictionaries is provided below. This function uses the convention that the bivector component corresponding to $e_2 \wedge e_3$ is associated with the first magnetic field component, and so forth.

```
def construct_em_field(e_field, b_field):
    """
    Construct the electromagnetic field bivector from the electric
    ↪ and magnetic field vectors.

    Parameters:
        e_field : dict
            Dictionary with keys 'e1', 'e2', 'e3' representing the
            ↪ electric field components.
        b_field : dict
```

```
            Dictionary with keys 'e1', 'e2', 'e3' representing the
         ↪   magnetic field components.

Returns:
    dict
        Dictionary representing the electromagnetic field as a
     ↪   multivector with the following keys:
            - 'e1', 'e2', 'e3' for the electric field components.
            - 'e23', 'e31', 'e12' for the magnetic field encoded
     ↪   as bivector components.

The construction is based on the relation:
    F = E1*e1 + E2*e2 + E3*e3 + B1*(e2^e3) + B2*(e3^e1) +
 ↪   B3*(e1^e2),
where the wedge product (^) expresses the formation of
 ↪   bivectors.
"""
F = {}
# Assign electric field components directly.
F['e1'] = e_field.get('e1', 0)
F['e2'] = e_field.get('e2', 0)
F['e3'] = e_field.get('e3', 0)
# Form the bivector components corresponding to the magnetic
 ↪   field.
F['e23'] = b_field.get('e1', 0)
F['e31'] = b_field.get('e2', 0)
F['e12'] = b_field.get('e3', 0)
return F
```

The function above encapsulates the construction of an electromagnetic field multivector. Input dictionaries for the electric and magnetic fields are combined to produce a complete representation that retains the distinct geometric nature of each field component. This explicitly layered structure supports rigorous numerical operations, such as the evaluation of the geometric derivative or the extraction of invariants.

Analysis of Electromagnetic Field Dynamics in Geometric Algebra

The analysis of electromagnetic phenomena in the framework of geometric algebra benefits from the clarity with which differential operations are performed on multivector fields. The operator ∇ acts on the electromagnetic field multivector F, and its geometric product decomposition naturally yields the divergence and curl in one unified operation. This methodological approach offers deeper insight into the symmetries of electromagnetism. For instance, the invariance properties associated with Lorentz transformations emerge directly from the algebraic structure.

In this context the field invariant

$$F^2 = F \cdot F$$

can be computed to extract physically significant quantities such as the difference between the squared magnitudes of the electric and magnetic fields. The representation of F within a single algebraic object permits an unambiguous determination of energy densities and facilitates the simulation of wave propagation in complex media. Moreover, the structured approach to assembling the electromagnetic field, as demonstrated by the construction function, ensures that numerical derivative operators and grade projection techniques generate results that preserve the underlying physical symmetries.

This algebraic perspective underpins robust simulation algorithms where the multivector nature of F is preserved at every computation step. The unified treatment of electric and magnetic fields within the geometric algebra framework streamlines both the symbolic manipulation and the numerical integration of Maxwell's equations, thereby enabling rigorous modeling and analysis of electromagnetic fields in a wide range of physical applications.

Python Code Snippet

```
import sympy as sp

# Define symbolic variables for spatial coordinates
x, y, z = sp.symbols('x y z')

# Define sample functions for the electric field components as
↪  functions of spatial variables
E1 = sp.sin(x)
E2 = sp.cos(y)
E3 = sp.exp(z)

# Define sample functions for the magnetic field components as
↪  functions of spatial variables
B1 = sp.cos(x)
B2 = sp.sin(y)
B3 = sp.log(1 + z)

# Electric field dictionary: corresponds to vector components
↪  (grade-1)
```

```python
e_field = {
    'e1': E1,
    'e2': E2,
    'e3': E3
}

# Magnetic field dictionary: In our GA representation, the magnetic
#    field components
# are mapped to the bivector parts. Here:
#   B1 corresponds to bivector (e2 ^ e3),
#   B2 corresponds to bivector (e3 ^ e1),
#   B3 corresponds to bivector (e1 ^ e2)
b_field = {
    'e1': B1,   # maps to bivector component 'e23'
    'e2': B2,   # maps to bivector component 'e31'
    'e3': B3    # maps to bivector component 'e12'
}

def construct_em_field(e_field, b_field):
    """
    Construct the electromagnetic field multivector from the
        electric and magnetic field components.

    The electromagnetic field F is represented as:
        F = E1*e1 + E2*e2 + E3*e3 + B1*(e2 ^ e3) + B2*(e3 ^ e1) +
            B3*(e1 ^ e2)

    Parameters:
    e_field: dict
        Dictionary with keys 'e1', 'e2', 'e3' for the electric
            field components.
    b_field: dict
        Dictionary with keys 'e1', 'e2', 'e3' representing the
            magnetic field components that
        map to bivector parts.

    Returns:
        A dictionary representing the electromagnetic field
            multivector with keys:
        'e1', 'e2', 'e3' for the vector (electric field)
            components,
        'e23', 'e31', 'e12' for the bivector (magnetic field)
            components.
    """
    F = {}
    # Assign electric field components directly (grade-1)
    F['e1'] = e_field.get('e1', 0)
    F['e2'] = e_field.get('e2', 0)
    F['e3'] = e_field.get('e3', 0)
    # Map magnetic field components to bivector parts (grade-2)
    F['e23'] = b_field.get('e1', 0)
    F['e31'] = b_field.get('e2', 0)
    F['e12'] = b_field.get('e3', 0)
```

```python
    return F

def compute_field_invariant(F):
    """
    Compute the electromagnetic field invariant, defined as:
      F~2 = F · F = |E|~2 - |B|~2
    where |E|~2 = E1~2 + E2~2 + E3~2 and |B|~2 = B1~2 + B2~2 + B3~2.

    In our multivector representation:
      Electric field components: F['e1'], F['e2'], F['e3']
      Magnetic field (bivector) components: F['e23'], F['e31'],
      ↪  F['e12']

    Returns:
       The simplified symbolic expression for F~2.
    """
    E_sq = F.get('e1', 0)**2 + F.get('e2', 0)**2 + F.get('e3', 0)**2
    B_sq = F.get('e23', 0)**2 + F.get('e31', 0)**2 + F.get('e12',
    ↪  0)**2
    invariant = sp.simplify(E_sq - B_sq)
    return invariant

def apply_nabla(F, var_symbols):
    """
    Apply the vector derivative operator (nabla) to the
    ↪  electromagnetic field F.

    The operator is defined as:
       nabla = e1 * /x + e2 * /y + e3 * /z
    For each component of F, we compute the derivative with respect
    ↪  to its corresponding spatial variable.

    Parameters:
       F: dict
          The electromagnetic field multivector dictionary.
       var_symbols: dict
          Dictionary with keys 'x', 'y', 'z' corresponding to the
          ↪  symbolic spatial variables.

    Returns:
       A dictionary representing the components of nabla F.
    """
    nabla_F = {}
    # Derivatives for the electric field components
    nabla_F['e1'] = sp.diff(F.get('e1', 0), var_symbols['x'])
    nabla_F['e2'] = sp.diff(F.get('e2', 0), var_symbols['y'])
    nabla_F['e3'] = sp.diff(F.get('e3', 0), var_symbols['z'])
    # Derivatives for the magnetic field bivector components (for
    ↪  demonstration)
    nabla_F['e23'] = sp.diff(F.get('e23', 0), var_symbols['x'])
    nabla_F['e31'] = sp.diff(F.get('e31', 0), var_symbols['y'])
    nabla_F['e12'] = sp.diff(F.get('e12', 0), var_symbols['z'])
    return nabla_F
```

```python
# Define a dictionary for the spatial derivative symbols
deriv_vars = {'x': x, 'y': y, 'z': z}

# Construct the electromagnetic field multivector using the provided
#   electric and magnetic fields
F = construct_em_field(e_field, b_field)

# Compute the electromagnetic field invariant (F^2 = |E|^2 - |B|^2)
field_invariant = compute_field_invariant(F)

# Apply the derivative operator (nabla) to F
nabla_F = apply_nabla(F, deriv_vars)

# Display the results
print("Electromagnetic Field Multivector F:")
for component, expression in F.items():
    print(f"{component} : {sp.pretty(expression)}")

print("\nField Invariant (F^2 = |E|^2 - |B|^2):")
sp.pprint(field_invariant)

print("\nGradient of F (nabla F) Components:")
for component, expression in nabla_F.items():
    print(f"{component} : {sp.pretty(expression)}")

if __name__ == '__main__':
    pass  # Execution demonstration; further integration and
          #   simulation can be added here.
```

Chapter 57

Relativistic Mechanics with GA: Computation in Spacetime

Relativistic Spacetime Algebra

Spacetime in relativistic mechanics is modeled as a four-dimensional manifold endowed with a Minkowski metric. In geometric algebra, a basis is adopted as $\{\gamma_0, \gamma_1, \gamma_2, \gamma_3\}$, where the metric signature is set so that $\gamma_0^2 = 1$ and $\gamma_i^2 = -1$ for $i = 1, 2, 3$. This construction allows the invariant interval to be compactly written as

$$x^2 = (x^0)^2 - (x^1)^2 - (x^2)^2 - (x^3)^2,$$

for a spacetime vector $x = x^0\gamma_0 + x^1\gamma_1 + x^2\gamma_2 + x^3\gamma_3$. The unified geometric product, combining both inner and outer products, encapsulates the invariant properties of such spacetime vectors and provides a powerful language to represent physical observables.

Lorentz Transformations and Boost Rotors

Within spacetime algebra, Lorentz transformations are implemented via rotors. A rotor is constructed as the exponential of a bivector that generates rotations and boosts within the spacetime plane. In the case of a boost along a spatial unit multivector n, the rotor R is expressed as

$$R = \exp\left(-\frac{\phi}{2}\gamma_0 n\right) = \cosh\left(\frac{\phi}{2}\right) - \gamma_0 n \sinh\left(\frac{\phi}{2}\right),$$

where ϕ is the rapidity parameter, related to the boost velocity by $\beta = \tanh(\phi)$. The transformation of any spacetime multivector M is achieved by the rotor sandwiching operation,

$$M' = R M \tilde{R},$$

with $\tilde{R}$ corresponding to the reverse of R. This rotor formulation directly guarantees the conservation of the spacetime interval and the covariance of the geometric relationships under Lorentz transformations.

Computational Implementation of Lorentz Boosts

The efficient simulation of spacetime dynamics demands an algorithmic construction of boost rotors. Computational implementations leverage standard routines for evaluating hyperbolic functions to translate the rapidity parameter into rotor components. A representative function encapsulates the procedure for computing the Lorentz boost rotor, taking as input the rapidity ϕ and a unit spatial multivector indicative of the boost direction.

```
def compute_boost_rotor(phi, spatial_direction):
    """
    Compute the Lorentz boost rotor for a boost along the specified
     spatial direction.

    The boost rotor R is defined as:
      R = exp(-phi/2 * gamma_0 * spatial_direction)
        = cosh(phi/2) - gamma_0 * spatial_direction * sinh(phi/2),
    where phi represents the rapidity parameter with beta =
     tanh(phi). The unit timelike
    basis vector gamma_0 and the unit spatial multivector
     spatial_direction must be defined
    within the global geometric algebra context.

    Parameters:
      phi : float
        The rapidity parameter.
```

```
spatial_direction : object
    A unit spatial multivector representing the boost
    ↪ direction.

Returns:
  object
      The computed Lorentz boost rotor.
"""
import math
cosh_half = math.cosh(phi / 2)
sinh_half = math.sinh(phi / 2)
rotor = cosh_half - (gamma_0 * spatial_direction) * sinh_half
return rotor
```

Spacetime Dynamics through GA Computational Operations

The rotor sandwiching operation provides a mechanism to perform Lorentz transformations on spacetime multivectors. When applied to energy–momentum vectors or field tensors, the transformation

$$M' = R M \tilde{R}$$

preserves invariant quantities, such as the spacetime interval, and yields covariant expressions suitable for numerical simulation. The structure of the geometric product, coupled with the proper management of multivector grades, enables discretized computations that track the evolution of relativistic systems with high fidelity.

This computational framework supports the derivation of invariant observables and facilitates the integration of derivative operations on multivector fields. The inherent parallelism and structured algebraic relationships simplify algorithms that address the challenges of simulating high-energy dynamics and gravitational interactions in a relativistically consistent manner.

Python Code Snippet

```
import math
from clifford import Cl

# Create a Minkowski spacetime geometric algebra with signature
↪ (1,3)
# Here, gamma0~2 = +1 and gamma1~2 = gamma2~2 = gamma3~2 = -1.
layout, blades = Cl(1, 3)
```

```
# Extract basis vectors
gamma0 = blades['e0']    # timelike basis vector
gamma1 = blades['e1']    # spacelike basis vector
gamma2 = blades['e2']    # spacelike basis vector
gamma3 = blades['e3']    # spacelike basis vector

def compute_boost_rotor(phi, spatial_direction):
    """
    Compute the Lorentz boost rotor for a boost along the specified
    ↪ spatial direction.

    Given the rapidity parameter phi and a unit spatial multivector
    ↪ representing
    the boost direction, the rotor R is defined as:

        R = exp(-phi/2 * gamma0 * spatial_direction)
          = cosh(phi/2) - gamma0 * spatial_direction * sinh(phi/2),

    where beta = tanh(phi) is the boost velocity parameter.

    Parameters:
      phi : float
          The rapidity parameter.
      spatial_direction : multivector
          A unit spatial multivector (e.g., gamma1, gamma2, or
          ↪ gamma3) defining the boost direction.

    Returns:
      rotor : multivector
          The computed Lorentz boost rotor.
    """
    cosh_half = math.cosh(phi / 2)
    sinh_half = math.sinh(phi / 2)
    rotor = cosh_half - (gamma0 * spatial_direction) * sinh_half
    return rotor

def apply_lorentz_transformation(multivector, rotor):
    """
    Apply a Lorentz transformation to a spacetime multivector using
    ↪ the rotor sandwich operation.

    The transformation is performed as:

        M' = R * M * ~R,

    where ~R is the reverse of the rotor R, ensuring the
    ↪ preservation of
    the spacetime interval.

    Parameters:
      multivector : multivector
```

            The spacetime multivector to be transformed (e.g., an
            ↪ event, energy-momentum vector, etc.).
        rotor : multivector
            The Lorentz boost rotor computed via
            ↪ compute_boost_rotor().

    Returns:
        transformed_mv : multivector
            The transformed multivector after applying the Lorentz
            ↪ boost.
    """
    return rotor * multivector * ~rotor

def main():
    # Define a sample spacetime vector:
    # Example: x = (t, x, y, z) = (1.0, 0.5, 0.0, 0.0)
    t, x, y, z = 1.0, 0.5, 0.0, 0.0
    spacetime_vector = t * gamma0 + x * gamma1 + y * gamma2 + z *
    ↪ gamma3

    # Define the boost: choose a boost along the gamma1 direction.
    # Since gamma1 is already a unit vector in the algebra, it
    ↪ represents the direction.
    boost_direction = gamma1

    # Choose a boost speed with beta = 0.5, so the rapidity phi is:
    beta = 0.5
    phi = math.atanh(beta)

    # Compute the boost rotor using the defined rapidity and
    ↪ direction.
    rotor = compute_boost_rotor(phi, boost_direction)

    # Transform the spacetime vector using the rotor sandwich
    ↪ operation.
    transformed_vector =
    ↪ apply_lorentz_transformation(spacetime_vector, rotor)

    # Output results for verification
    print("Original spacetime vector:")
    print(spacetime_vector)
    print("\nBoost rotor (R):")
    print(rotor)
    print("\nTransformed spacetime vector (R M ~R):")
    print(transformed_vector)

if __name__ == "__main__":
    main()

Chapter 58

Computational Implementations: Designing GA Algorithms

Foundations of Multivector Representation

The design of software algorithms for geometric algebra (GA) is underpinned by the choice of data structures that represent multivectors. In many physics applications, a multivector is expressed in the form

$$M = \sum_A m_A\, e_A,$$

where the index set A corresponds to the ordered basis blades of the algebra and the coefficients m_A are typically stored as floating point numbers. One common approach is to use dictionary structures in which the keys are tuples identifying the basis blades and the values are the corresponding numerical coefficients. This representation can be highly efficient when the multivectors are sparse, and it permits flexibility in managing algebraic operations that involve varying grades.

The data structure selection influences not only memory usage but also the computational overhead when iterating through the non-zero components during operations such as addition, inversion, or the geometric product. In many implementations, additional precomputation of lookup tables for blade multiplications is utilized to shorten runtime for repetitive product calculations in high-dimensional algebras.

Algorithmic Implementation of the Geometric Product

Among all GA operations, the geometric product is the most fundamental, as it encapsulates both the inner and outer products. Its algorithmic implementation requires careful management of the combinatorial aspects of basis blade multiplication, particularly the ordering of indices and the impact of the metric signature on the resulting sign. The procedure typically involves iterating over all pairs of basis components in two multivectors and merging their associated index tuples. Every combination contributes a factor determined by the coefficients and the sign adjustment emerging from the anti-commutative properties of the basis vectors.

The conceptual steps in the algorithm can be summarized as follows:

- Each multivector is stored as a mapping from a tuple (representing a basis blade) to a coefficient.

- For every pair from the two input multivectors, the corresponding basis blades are concatenated to form a new blade.

- The new blade is normalized by sorting its indices, and a sign factor is computed based on the permutations needed to reach the sorted order.

- The coefficients of matching blades are accumulated, yielding the final multivector product.

A function implementing this algorithm is presented below. The code snippet illustrates the process within a Python function while commenting on critical steps, including a simplified approach to compute the sign factor based on the grades of the individual blades.

```
def geometric_product(mv1, mv2):
    """
```

```
Compute the geometric product of two multivectors represented as
↪ dictionaries.

Each multivector is expressed as a dictionary where keys are
↪ tuples of integers,
representing the indices of the basis blades, and the values are
↪ floating-point
coefficients corresponding to these blades.

The algorithm iterates over each pair of blades from mv1 and
↪ mv2, combines the index
tuples, and sorts the combined tuple to account for the
↪ reordering required by the
anti-commutative properties of the basis vectors. A simplified
↪ sign determination
is applied based on the product of the grades of the individual
↪ blades.

Parameters:
  mv1 : dict
      The first multivector, with keys indicating basis blades.
  mv2 : dict
      The second multivector, following the same representation.

Returns:
  dict
      A dictionary representing the geometric product of mv1 and
      ↪ mv2.
"""
result = {}
for blade1, coeff1 in mv1.items():
    for blade2, coeff2 in mv2.items():
        combined_blade = blade1 + blade2
        sorted_blade = tuple(sorted(combined_blade))
        # The sign factor here is computed in a simplified
        ↪ manner based on the
        # lengths of the blades; a complete implementation would
        ↪ require a
        # permutation parity function that precisely captures
        ↪ the effect of
        # reordering the basis vectors.
        sign = (-1) ** (len(blade1) * len(blade2))
        prod_coeff = coeff1 * coeff2 * sign
        result[sorted_blade] = result.get(sorted_blade, 0) +
        ↪ prod_coeff
return result
```

The above function embodies the core steps of a geometric product computation and illustrates how the combination of data structure design and algorithmic optimization yields a viable implementation for practical applications in physics.

Runtime Efficiency and Precision Management

A consequential factor in the design of GA algorithms is the runtime efficiency achieved through data structure optimization and algorithmic fine-tuning. In large-scale physical simulations, a multivector might possess a high dimensionality with many non-zero components. In these contexts, clustering operations, caching precomputed multiplications, and using sparse representations become essential. Optimizing the inner loops of the geometric product calculation contributes significantly to overall performance.

Precision management is equally critical, especially when dealing with floating-point arithmetic in high-energy or relativistic contexts. The inherent error propagation in iterative computations necessitates careful control of numerical precision. In some cases, symbolic computation libraries may be integrated with numerical routines to verify critical intermediate results.

Furthermore, the choice between symbolic and numerical methods is influenced by the nature of the physical problem. For dynamical simulations, where iterative time-stepping is common, the use of optimized numerical routines can lead to substantial performance gains. Meanwhile, symbolic methods are employed for analytical derivations that validate the structure of the computational algorithms.

The interleaving of theoretical insights from physics with practical computer science considerations thus informs the development of robust GA software libraries capable of handling the complexities inherent in modeling physical phenomena with multi-graded algebraic entities.

Python Code Snippet

```
# This module implements key operations in Geometric Algebra (GA)
    using a dictionary-based
# multivector representation. Each multivector is a dictionary where
    keys are tuples
# representing basis blades (e.g., (1,2) stands for ee) and values
    are floating-point
# coefficients. The Euclidean metric is assumed by default, although
    a custom metric can be
# provided as a dictionary mapping basis vector indices to their
    square values.
```

```
#
# The following functions are included:
#   - permutation_sign: Computes the sign of a permutation for a
↪     (possibly non-distinct) list.
#   - canonical_blade: Transforms a blade (possibly unsorted and
↪     with duplicate indices)
#                     into its canonical form by sorting and
↪     cancelling duplicate pairs.
#   - geometric_product: Computes the geometric product of two
↪     multivectors.
#   - add_multivectors: Adds two multivectors.
#   - print_multivector: Nicely prints a multivector.
#
# An example is provided in the main block to illustrate usage.

def permutation_sign(seq):
    """
    Compute the sign of a permutation for a sequence with possible
    ↪   duplicate elements.
    This function counts inversions between pairs of elements that
    ↪   are strictly greater,
    and returns (-1)^(inversion_count).
    """
    inversion_count = 0
    n = len(seq)
    for i in range(n):
        for j in range(i + 1, n):
            # Only count inversion if elements are strictly
            ↪   out-of-order.
            if seq[i] > seq[j]:
                inversion_count += 1
    return (-1) ** inversion_count

def canonical_blade(blade, metric=None):
    """
    Convert a blade (given as a tuple of basis indices, possibly
    ↪   unsorted or with duplicates)
    to its canonical form by:
      - Sorting the indices while computing the sign factor due to
      ↪   reordering.
      - Cancelling pairs of identical indices according to the
      ↪   metric.

    In geometric algebra, for an orthonormal basis:
       e_i * e_i = metric[i], typically 1 (Euclidean) or -1 (for
       ↪   some signatures).
    Duplicate indices are handled by reducing them in pairs.

    Parameters:
      blade : tuple
          A tuple of integers representing the basis vectors (e.g.,
          ↪   (3,1,2,1)).
      metric : dict or None
```

```
    A dictionary mapping basis indices to their square value.
    If None, it is assumed that metric[i] = 1 for all i.

Returns:
  (coeff_factor, canon_blade) : (float, tuple)
      coeff_factor: the accumulated scalar (from both sign
      ↪ changes and metric entries).
      canon_blade: the canonical blade represented as a sorted
      ↪ tuple with duplicates removed
                   (if an index appears an even number of times,
                   ↪ it cancels out).
"""
if metric is None:
    metric = {}

# Helper: get metric value, defaulting to 1 for Euclidean
↪ metric.
def m_val(i):
    return metric.get(i, 1)

# Convert blade to list for processing.
blade_list = list(blade)
# Compute sign factor from the reordering required to sort the
↪ blade.
sorted_list = sorted(blade_list)
sign = permutation_sign(blade_list)

# Count frequency of each index.
counts = {}
for i in sorted_list:
    counts[i] = counts.get(i, 0) + 1

coeff_factor = sign
canon = []
# Process each index in sorted order; pair cancellations
↪ contribute metric factors.
for idx in sorted(counts.keys()):
    count = counts[idx]
    pairs = count // 2
    coeff_factor *= (m_val(idx) ** pairs)
    # For an odd count, one copy remains.
    if count % 2 == 1:
        canon.append(idx)
canon_blade = tuple(canon)
return coeff_factor, canon_blade

def geometric_product(mv1, mv2, metric=None):
    """
    Compute the geometric product of two multivectors.

    Each multivector is represented as a dictionary with:
      - Keys: tuples representing basis blades (e.g., () for scalar,
      ↪ (1,) for e, (2,3) for ee).
```

        - Values: floating-point coefficients.

    The product is calculated by iterating over each pair of basis
    ↪ blades from mv1 and mv2,
    concatenating their index tuples, and then converting the
    ↪ combined blade into its canonical
    form using the canonical_blade function (which handles
    ↪ reordering and cancellation).

    Parameters:
      mv1 : dict
          A multivector, for example: {(): 1.0, (1,): 2.0, (2,3):
          ↪ 3.0}
      mv2 : dict
          Another multivector with the same representation.
      metric : dict or None
          A dictionary specifying the metric; if None, Euclidean
          ↪ metric is assumed.

    Returns:
      dict
          The resulting multivector from the geometric product.
    """
    result = {}
    for blade1, coeff1 in mv1.items():
        for blade2, coeff2 in mv2.items():
            # Concatenate the basis indices from both blades.
            combined = blade1 + blade2
            # Get the canonical form: a scalar factor (from sign and
            ↪ cancellations)
            # and the corresponding sorted blade.
            gp_coeff, canon_blade = canonical_blade(combined,
            ↪ metric)
            prod_coeff = coeff1 * coeff2 * gp_coeff
            # Accumulate coefficients for matching blades.
            result[canon_blade] = result.get(canon_blade, 0) +
            ↪ prod_coeff
    return result

def add_multivectors(mv1, mv2):
    """
    Add two multivectors represented as dictionaries.

    Parameters:
      mv1, mv2 : dict
          Multivectors to be added.

    Returns:
      dict
          A new multivector representing the sum.
    """
    result = mv1.copy()
    for blade, coeff in mv2.items():
```
```

```python
            result[blade] = result.get(blade, 0) + coeff
    return result

def print_multivector(mv):
    """
    Print a multivector in a human-readable format.

    The output shows scalar, vector, and higher-grade terms in a
    ↪ concise representation.

    Parameters:
      mv : dict
          A multivector represented as a dictionary.
    """
    terms = []
    # Sorting primarily by grade (length of blade) then
    ↪ lexicographically.
    for blade, coeff in sorted(mv.items(), key=lambda x: (len(x[0]),
    ↪ x[0])):
        # Represent the scalar part.
        if blade == ():
            terms.append(f"{coeff:.3g}")
        else:
            # Format the blade as e<number>.
            blade_str = "".join(f"e{idx}" for idx in blade)
            terms.append(f"{coeff:.3g}{blade_str}")
    print(" + ".join(terms) if terms else "0")

# Example usage demonstrating the GA operations.
if __name__ == "__main__":
    # Define a metric for an orthonormal Euclidean space (default:
    ↪ metric[i]=1).
    metric = {1: 1, 2: 1, 3: 1}

    # Define two multivectors:
    # mv1 represents: 1 + 2e + 3(ee)
    mv1 = {
        (): 1.0,
        (1,): 2.0,
        (2, 3): 3.0
    }

    # mv2 represents: 4 + 5e + 6(ee)
    mv2 = {
        (): 4.0,
        (2,): 5.0,
        (1, 3): 6.0
    }

    print("Multivector 1:")
    print_multivector(mv1)

    print("\nMultivector 2:")
```

```
print_multivector(mv2)

# Compute the geometric product of mv1 and mv2.
gp = geometric_product(mv1, mv2, metric)

print("\nGeometric Product (mv1 * mv2):")
print_multivector(gp)
```

Chapter 59

Optimizing the Geometric Product: Efficiency in Computation

Precomputation Strategies

In geometric algebra, the geometric product of two multivectors,

$$M = \sum_A m_A\, e_A, \quad N = \sum_B n_B\, e_B,$$

is computed by processing a large number of basis blade pairs. Since each product $e_A e_B$ adheres to deterministic sign conventions and cancellation rules, caching intermediate results is essential. Precomputation techniques reduce redundant recomputation of permutation factors and canonical blade representations. A precomputed lookup table may store the product of two basis blades once computed, thereby reducing the impact of nested iterations in high-dimensional applications.

A representative function that implements a caching strategy for the product of two basis blades is provided below.

```
def optimized_blade_product(blade1, blade2, metric, cache):
    """
```

```
Compute the geometric product of two basis blades using a
↪ precomputed cache.

Parameters:
    blade1 : tuple of int
        Represents the indices in the first basis blade.
    blade2 : tuple of int
        Represents the indices in the second basis blade.
    metric : dict
        A dictionary mapping basis vector indices to their squared
        ↪ norm.
    cache : dict
        A lookup table storing previously computed products for
        ↪ pairs of blades.

Returns:
    (factor, canonical_blade) : (float, tuple)
        'factor' is the overall scalar multiplication factor
        ↪ resulting from sign changes
        and metric contributions. 'canonical_blade' is the ordered
        ↪ tuple representing the
        resulting blade after cancellation of repeated indices.
"""
key = (blade1, blade2)
if key in cache:
    return cache[key]
# Compute the product (omitted: detailed permutation and
↪ cancellation logic)
factor = 1.0  # placeholder for computed scalar factor
combined = blade1 + blade2
canonical_blade = tuple(sorted(combined))   # naive ordering,
↪ without cancellation
cache[key] = (factor, canonical_blade)
return factor, canonical_blade
```

This function illustrates the central idea behind precomputation: the result of the operation on a specific pair of basis blades is stored and reused, thereby amortizing the cost of expensive operations such as sorting and sign determination across many iterations.

Data Structure Optimization

Efficient representation of multivectors is a cornerstone for high-performance computation. Multivectors are typically stored as mappings from basis blades to their corresponding coefficients. When the blades are represented as tuples of integers, dictionary structures in programming languages such as Python are well suited for sparsity and rapid look-up. However, the cost of tuple manipulation—particularly concatenation and sorting—tends to be non-negligible in inner loops.

Alternative representations, such as bitmask encodings, have been explored to represent blades compactly. In these schemes, each basis blade is associated with an integer where each bit indicates the presence or absence of a corresponding basis vector. Such representations allow the replacement of tuple sorting by fast bitwise operations. Furthermore, maintaining compact data structures improves memory access patterns and cache locality, which is critical in computationally intensive applications.

Meticulous profiling of memory usage and access speeds yields insights that guide the selection and refinement of underlying data structures, ensuring that the overhead associated with algebraic manipulations is minimized.

Algorithmic Enhancements

The efficient computation of the geometric product hinges on reducing the complexity inherent in its iterative evaluation. When processing products of multivectors,

$$MN = \sum_{A,B} m_A n_B \, (e_A e_B),$$

the number of operations grows rapidly with the number of nonzero components. Algorithmic enhancements focus on merging operations, applying loop unrolling, and taking advantage of the anti-commutative properties of basis vectors to combine like terms early in the computation.

Techniques such as vectorization and parallelism are employed where the underlying hardware supports such operations. Batch processing of blade pairs can be implemented to leverage modern CPU vector instructions or GPU parallelism. In addition, the exploitation of algebraic symmetries allows the elimination of redundant evaluations, with precomputed lookup tables serving as an effective bridge between theoretical formulation and practical implementation.

These optimizations are achieved by carefully balancing precomputation overhead with runtime benefits during iterative simulations. Profiling and benchmarking of optimized routines reveal substantial improvements in scenarios with high blade multiplicities, validating the approaches discussed herein.

Python Code Snippet

```
"""
Comprehensive Python implementation for optimizing the geometric
↪ product in Geometric Algebra.
This code snippet demonstrates:
  - Precomputation with caching of basis blade products.
  - Data structure optimizations for multivector representation.
  - Algorithmic enhancements such as early cancellation and sign
    ↪ tracking during blade multiplication.
  - Optional bitmask conversion for compact blade representation.
"""

def optimized_blade_product(blade1, blade2, metric, cache):
    """
    Compute the geometric product of two basis blades using a
    ↪ precomputed cache.

    Parameters:
      blade1 : tuple of int
          Sorted tuple representing the indices in the first basis
          ↪ blade (e.g., (1, 3)).
      blade2 : tuple of int
          Sorted tuple representing the indices in the second basis
          ↪ blade (e.g., (2, 3)).
      metric : dict
          A dictionary mapping basis vector indices to their squared
          ↪ norm
          (e.g., {1: 1, 2: 1, 3: 1} for a Euclidean metric).
      cache : dict
          Lookup table storing previously computed products for
          ↪ pairs of blades.

    Returns:
      (factor, canonical_blade) : (float, tuple)
          'factor' is the overall scalar coefficient resulting from
          ↪ sign changes
          and metric contributions during cancellation.
          'canonical_blade' is the ordered tuple representing the
          ↪ resulting blade after cancellation.
    """
    key = (blade1, blade2)
    if key in cache:
        return cache[key]

    # Combine the indices from both blades.
    combined = list(blade1) + list(blade2)

    # Compute the sign factor: count the number of swaps required to
    ↪ sort the combined list.
    sign = 1
    # Using a simple bubble sort algorithm for clarity.
```

```
            for i in range(len(combined)):
                for j in range(i + 1, len(combined)):
                    if combined[i] > combined[j]:
                        combined[i], combined[j] = combined[j], combined[i]
                        sign *= -1  # Each swap flips the sign.

            # Perform cancellation: if an index appears consecutively, they
            ↪ cancel subject to metric.
            canonical_blade = []
            i = 0
            while i < len(combined):
                if (i + 1) < len(combined) and combined[i] == combined[i +
                ↪ 1]:
                    # Cancel the two identical basis vectors; multiply in
                    ↪ the metric contribution.
                    sign *= metric.get(combined[i], 1)
                    i += 2  # Skip the cancelled pair.
                else:
                    canonical_blade.append(combined[i])
                    i += 1
            canonical_blade = tuple(canonical_blade)

            cache[key] = (sign, canonical_blade)
            return sign, canonical_blade

     def geometric_product(multivector1, multivector2, metric):
         """
         Compute the geometric product of two multivectors.

         Each multivector is represented as a dictionary mapping a basis
         ↪ blade (tuple) to its coefficient.
         The scalar is represented by the empty tuple ().

         Parameters:
           multivector1 : dict
               The first multivector (e.g., {(): 3.0, (1,): 2.0, (2,3):
               ↪ 1.0}).
           multivector2 : dict
               The second multivector (e.g., {(): 1.0, (1,): 1.0}).
           metric : dict
               Metric dictionary mapping basis vector indices to their
               ↪ squared norms.

         Returns:
           result : dict
               A dictionary representing the resulting multivector from
               ↪ the geometric product.
         """
         result = {}
         cache = {}  # Cache to amortize the cost of computing blade
         ↪ products.

         # Iterate over every pair of blades from both multivectors.
```

```
        for blade1, coeff1 in multivector1.items():
            for blade2, coeff2 in multivector2.items():
                factor, canonical_blade =
                ↪   optimized_blade_product(blade1, blade2, metric,
                ↪   cache)
                new_coeff = coeff1 * coeff2 * factor
                if canonical_blade in result:
                    result[canonical_blade] += new_coeff
                else:
                    result[canonical_blade] = new_coeff
        return result

def bitmask_representation(blade, n):
    """
    Optional: Convert a basis blade represented as a tuple to a
    ↪   bitmask integer.

    Parameters:
      blade : tuple of int
          Sorted tuple representing basis blade indices.
      n : int
          Total number of basis vectors in the space.

    Returns:
      bitmask : int
          Integer with bits set corresponding to the presence of
          ↪   basis vectors.
    """
    bitmask = 0
    for index in blade:
        # Set the (index-1)th bit (assuming basis vectors are
        ↪   1-indexed).
        bitmask |= (1 << (index - 1))
    return bitmask

if __name__ == "__main__":
    # Define a Euclidean metric for a 3-dimensional space.
    metric = {1: 1, 2: 1, 3: 1}

    # Define multivectors using dictionary representations:
    # Multivector A = 3 + 2e1 + e2e3, where the scalar is given by
    ↪   ().
    A = {(): 3.0, (1,): 2.0, (2, 3): 1.0}
    # Multivector B = 1 + e1.
    B = {(): 1.0, (1,): 1.0}

    # Compute the geometric product of A and B.
    product = geometric_product(A, B, metric)

    print("Geometric Product of A and B:")
    for blade, coeff in product.items():
        # Optionally, convert the blade representation to a bitmask
        ↪   for compact display.
```

```
bitmask = bitmask_representation(blade, 3) if blade else 0
print("Blade:", blade, "Bitmask:", bin(bitmask),
↪    "Coefficient:", coeff)
```

Chapter 60

Numerical Methods in GA: Tackling Complex Problems

Foundations of Numerical Computation in Geometric Algebra

The numerical treatment of geometric algebra necessitates the development of frameworks that can accurately represent and manipulate multivector quantities. Each multivector is composed of a weighted sum of basis blades, where the blades are defined by ordered sets of indices corresponding to the geometric directions in the underlying space. When represented in a numerical environment, these objects often adopt a sparse mapping structure that associates each blade with its corresponding coefficient. The exponential growth of the number of nonzero components with increasing dimensionality imposes critical demands on memory management and computational efficiency. In this context, careful attention is required to preserve the algebraic structure, particularly during operations such as the geometric product, which involves complex interactions dictated by metric signatures and permutation sign conventions. The precise encoding of these properties is essential to maintain the fidelity of physical simulations.

Discretization and Representation Schemes

There exist several discretization techniques for transferring continuously defined multivector fields into a computationally tractable form. Finite difference and finite element methods have been adapted to account for the composite structure of multivectors. In these methods, the continuous domain is partitioned into a grid or mesh, and the field values are approximated by the multivector components sampled at discrete points. Special quadrature rules are employed to ensure that the integrals over these discrete elements accurately integrate the contributions from different grades. The projection of continuous differential operators onto a discrete basis results in algebraic systems where the differential relationships are reinterpreted as operations among the multivector components. This procedure preserves the interference and cancellation effects intrinsic to geometric algebra, as observed in the handling of the wedge product and its associated cancellations.

Iterative Solvers and Convergence Techniques

Numerical solutions of systems formulated in the geometric algebra framework often involve iterative methods that are designed to handle the multivector nature of the unknowns. Iterative solvers, including methods such as conjugate gradient and generalized minimal residual (GMRES), are extended to operate on systems where each equation is endowed with a multivector structure. In these scenarios, convergence is measured in terms of a norm defined from the scalar product intrinsic to the algebra, typically expressed as

$$\|M\| = \sqrt{\langle M^2 \rangle_0},$$

where $\langle M^2 \rangle_0$ denotes the scalar part of the squared multivector. Special attention is given to maintaining numerical stability during each iteration, as the complex interplay of sign changes and metric contractions can lead to nontrivial error propagation. Adaptive strategies are often necessary, involving dynamically adjusted relaxation factors and convergence thresholds that cater to the nonlinearity and high-dimensional character of the system.

Adaptive Integration Methods for Multivector Fields

The integration of differential equations where the derivatives are represented within the geometric algebra framework demands adaptive techniques that can accommodate both the local nuances of the field evolution and the global geometric structure. High-order Runge–Kutta methods, modified to handle multivector inputs and outputs, are employed to ensure that the discretized integration accurately captures the evolution of the system. The adaptation of standard integration procedures involves error estimation techniques that compare intermediate approximations derived from different orders of the numerical scheme. This allows for the dynamic adjustment of the integration step size in order to minimize local truncation error while preserving discrete invariants. The adaptive techniques are essential in situations where the field variables exhibit rapid variations or where the underlying geometric constraints impose stiff behavior on the numerical solution.

Error Propagation and Stability Analysis

The propagation of numerical error in iterative methods applied to geometric algebra is a critical concern, particularly when computations involve a sequence of geometric products that are sensitive to rounding and cancellation effects. An in-depth stability analysis is required to ensure that the overall error magnitude does not grow uncontrollably during the simulation. The interaction between the sign alternations introduced by the permutation of indices and the metric factors can lead to non-intuitive pathways for error amplification. Detailed evaluations consider the evolution of the residual error in each iterative update, taking into account the multivector norm

$$\|M\| = \sqrt{\langle M^2 \rangle_0},$$

and its associated contraction properties. Analytical bounds for error propagation are established by rigorously examining the influence of derivative discretizations and iterative approximations. The stability criteria are designed to guarantee that the numerical solution faithfully reproduces the theoretical behavior of the physical system modeled by the geometric algebra framework.

Python Code Snippet

```python
import math

# Global metric for Euclidean space (example: 4-dimensional)
# Here, the metric is defined such that for each basis vector e_i,
#   e_i*e_i = 1.
METRIC = {1: 1, 2: 1, 3: 1, 4: 1}


def geometric_product_blades(blade1, blade2, metric):
    """
    Compute the geometric product of two basis blades given as
        tuples.

    The product is computed by concatenating the two blades, then
    reordering them to canonical order while keeping track of the
        parity (sign)
    of the permutation. Duplicate indices are removed with an
        accompanying
    metric factor.
    """
    # Combine the blades into a single list.
    result = list(blade1) + list(blade2)
    sign = 1
    # Bubble sort to order the indices and count swaps (each swap
    #   multiplies by -1).
    n = len(result)
    for i in range(n):
        for j in range(i + 1, n):
            if result[i] > result[j]:
                sign = -sign
                result[i], result[j] = result[j], result[i]
    # Remove duplicate indices applying the metric.
    final_blade = []
    i = 0
    while i < len(result):
        if i + 1 < len(result) and result[i] == result[i + 1]:
            # When the same index appears twice, they cancel to the
            #   metric factor.
            sign *= metric.get(result[i], 1)
            i += 2  # Skip the duplicate pair.
        else:
            final_blade.append(result[i])
            i += 1
    return tuple(final_blade), sign


class Multivector:
    """
    A simple implementation of a multivector in geometric algebra.
```

```
Each multivector is represented as a dictionary mapping a basis
↪   blade
(expressed as a tuple of integers; the scalar is represented as
↪   an empty tuple)
to its corresponding coefficient.
"""
def __init__(self, components):
    # components: dict where keys are tuples (e.g., () for
    ↪   scalar, (1,) for e1, (1,2) for e1e2, etc.)
    self.components = components.copy()

def __add__(self, other):
    result = self.components.copy()
    for blade, coeff in other.components.items():
        result[blade] = result.get(blade, 0) + coeff
    return Multivector(result)

def __sub__(self, other):
    result = self.components.copy()
    for blade, coeff in other.components.items():
        result[blade] = result.get(blade, 0) - coeff
    return Multivector(result)

def __rmul__(self, scalar):
    # Scalar multiplication from the left.
    return Multivector({blade: scalar * coeff for blade, coeff
    ↪   in self.components.items()})

def __mul__(self, other):
    """
    Geometric product between two multivectors.
    For each pair of basis blades, compute their product using
    ↪   the metric.
    """
    if isinstance(other, Multivector):
        result = Multivector({})
        for blade1, coeff1 in self.components.items():
            for blade2, coeff2 in other.components.items():
                new_blade, sign =
                ↪   geometric_product_blades(blade1, blade2,
                ↪   METRIC)
                result.components[new_blade] =
                ↪   result.components.get(new_blade, 0) + coeff1
                ↪   * coeff2 * sign
        return result
    else:
        # Allow multiplication by a scalar from the right.
        return self.__rmul__(other)

def scalar(self):
    """Return the scalar part (grade 0) of the multivector."""
    return self.components.get((), 0)
```

```python
def norm(self):
    """
    Compute the norm of the multivector defined as:
        ||M|| = sqrt( ( M*M )_0 )
    where <·>_0 extracts the scalar part.
    """
    prod = self * self
    s = prod.scalar()
    # In some cases the scalar part might be negative, so take
    ↪  the absolute value.
    return math.sqrt(abs(s))

def __str__(self):
    # Provide an informative string representation.
    terms = []
    for blade in sorted(self.components, key=lambda b: (len(b),
    ↪  b)):
        coeff = self.components[blade]
        terms.append(f"{coeff:+.3f}e{blade}")
    return " ".join(terms)


def inner_product(M, N):
    """
    Define an inner product for multivectors as the scalar part of
    ↪  the
    geometric product.
    """
    return (M * N).scalar()


def conjugate_gradient(A, b, x0, tol=1e-6, max_iter=100):
    """
    Conjugate Gradient solver adapted for a system in geometric
    ↪  algebra:
        A(x) = b
    where A is a linear operator on multivectors.
    The norm used is:
        ||M|| = sqrt( < M*M >_0 ).
    """
    x = x0
    r = b - A(x)
    p = r
    rsold = inner_product(r, r)
    for i in range(max_iter):
        Ap = A(p)
        alpha = rsold / inner_product(p, Ap)
        x = x + alpha * p
        r = r - alpha * Ap
        rsnew = inner_product(r, r)
        if math.sqrt(rsnew) < tol:
            print(f"Conjugate Gradient converged after {i+1}
            ↪  iterations with residual {math.sqrt(rsnew):.3e}")
```

```
        return x
    beta = rsnew / rsold
    p = r + beta * p
    rsold = rsnew
print("Conjugate Gradient did not converge within the maximum
    ↪ number of iterations.")
return x


def runge_kutta4(f, x0, t0, t_end, dt):
    """
    Standard 4th order Runge-Kutta integrator adapted for ODEs on
    ↪ multivector fields.
    It integrates the ODE:
        dx/dt = f(t, x)
    from t0 to t_end with step size dt.
    """
    t = t0
    x = x0
    steps = int((t_end - t0) / dt)
    for _ in range(steps):
        k1 = f(t, x)
        k2 = f(t + dt/2, x + (dt/2) * k1)
        k3 = f(t + dt/2, x + (dt/2) * k2)
        k4 = f(t + dt, x + dt * k3)
        x = x + (dt/6) * (k1 + 2*k2 + 2*k3 + k4)
        t += dt
    return x


# Example linear operator for use with the Conjugate Gradient
↪ solver.
def A_operator(x):
    """
    Define a sample linear operator on multivectors.
    For demonstration, we define:
        A(x) = x + 0.5 * (e1 * x)
    where e1 is the basis vector represented by (1,).
    """
    e1 = Multivector({(1,): 1})
    return x + 0.5 * (e1 * x)


if __name__ == '__main__':
    # Example usage of the multivector operations and numerical
    ↪ algorithms.

    # Define two sample multivectors.
    M = Multivector({(): 2.0, (1,): 3.0, (2, 3): -1.0})
    N = Multivector({(): 1.0, (1, 2): 4.0})

    # Compute and display the geometric product.
    product = M * N
```

```
print("Geometric Product M * N =", product)
print("Norm of M =", M.norm())

# Solve a linear system A(x) = b using the Conjugate Gradient
↪   method.
b = Multivector({(): 5.0, (1,): 1.0})
x0 = Multivector({})  # Zero multivector as the initial guess.
x_sol = conjugate_gradient(A_operator, b, x0)
print("Solution x for A*x = b:", x_sol)

# Solve a differential equation using the Runge-Kutta 4 method.
# Consider the ODE: dx/dt = a * x with a constant generator a.
a = Multivector({(2,): 1.0})
def ode_func(t, x):
    return a * x  # Right-hand side of the ODE.

x0_rk = Multivector({(): 1.0})  # Initial condition.
x_end = runge_kutta4(ode_func, x0_rk, t0=0, t_end=1.0, dt=0.01)
print("Runge-Kutta integration result x(1):", x_end)
```

Chapter 61

Symbolic Algorithms: Automating GA Computations

Foundations of Symbolic Computation in Geometric Algebra

Geometric algebra elements can be represented symbolically as combinations of basis blades with coefficients that are algebraic expressions. Multivectors, in this context, are decomposed into sums of components where each term corresponds to a blade defined by an ordered set of indices. The geometric product is encoded symbolically by explicitly stating the permutation rules, incorporating sign factors that account for the reordering of indices, and applying metric contractions for repeated indices. This symbolic formalism enables a systematic treatment of non-commutative products and the inherent combinatorial complexity that arises in high-dimensional spaces. By decomposing expressions into their fundamental symbolic constituents, automated systems can apply pattern matching and rewriting rules to enforce canonical forms and facilitate further manipulation.

Automated Simplification Techniques in Geometric Algebra

The automated simplification of symbolic expressions in geometric algebra is achieved through a sequence of algorithmic rewriting rules. These rules standardize the representation of multivectors by reordering the indices in each basis blade, eliminating duplicate indices according to the metric rules, and aggregating coefficients for like terms. Through recursive application of these procedures, the system transforms complex expressions into a form that is compatible with the algebraic structure of geometric algebra. The simplification process is instrumental in reducing the length and complexity of expressions before further analytical or numerical treatment.

An essential component of this process is the canonical reordering of basis blades. The following function demonstrates a method for reordering a basis blade symbolically. The function returns the blade in ascending order along with a sign that reflects the parity of the permutation. This method ensures that the reordering respects the anti-symmetric properties of the geometric product and provides a reliable building block for further symbolic manipulations.

```
def symbolic_reorder_blade(blade):
    """
    Reorder indices of a blade symbolically for geometric algebra.

    Parameters:
        blade : tuple
            A tuple representing a basis blade, for instance (3, 1,
            ↪ 2).

    Returns:
        tuple
            A tuple (canonical_blade, sign) where canonical_blade is
            ↪ the indices sorted in ascending order
            and sign is either +1 or -1 reflecting the number of
            ↪ swaps made.

    The function employs a bubble sort algorithm to reorder the
    ↪ indices. Each swap introduces a factor of -1,
    ensuring that the overall sign correctly represents the
    ↪ permutation parity required to achieve the canonical order.
    """
    canonical_blade = list(blade)
    sign = 1
```

```
n = len(canonical_blade)
for i in range(n):
    for j in range(0, n - i - 1):
        if canonical_blade[j] > canonical_blade[j + 1]:
            canonical_blade[j], canonical_blade[j + 1] =
             ↪ canonical_blade[j + 1], canonical_blade[j]
            sign = -sign
return tuple(canonical_blade), sign
```

Algorithmic Implementation and Efficiency Considerations

The computational implementation of symbolic algorithms in geometric algebra not only requires correctness in applying algebraic rules but also demands rigorous attention to efficiency. Data structures are crafted to exploit the inherent sparsity of multivector representations while permitting rapid access to and modification of individual blade components. Hash maps and tree-based representations are commonly utilized to index terms based on their blade structure. Furthermore, the precomputation and caching of frequently encountered sub-expressions contribute to a significant reduction in redundant calculations during the simplification process.

Algorithmic efficiency is also impacted by the design of the symbolic rewriting system. Techniques derived from term rewriting systems and pattern matching algorithms ensure that transformations are applied only when necessary, thereby controlling the exponential growth in complexity typically associated with geometric products in high-dimensional settings. Adaptive methods for error control and intermediate result caching are integrated into the symbolic engine to maintain both precision and speed. This careful balance between algorithmic generality and specialized implementation strategies is critical to automating symbolic computations while preserving the structural integrity of geometric algebra.

Python Code Snippet

```
# Define the metric for our geometric algebra.
# For simplicity, we assume an Euclidean metric where each basis
 ↪ vector squares to +1.
# Extend or modify this dictionary to accommodate different
 ↪ signatures.
```

```python
metric = {
    1: 1,
    2: 1,
    3: 1,
    4: 1  # Extend as needed for higher dimensions.
}

def symbolic_reorder_blade(blade):
    """
    Reorder indices of a blade symbolically for geometric algebra.

    Parameters:
        blade : tuple
            A tuple representing a basis blade, e.g., (3, 1, 2).

    Returns:
        tuple:
            A tuple (canonical_blade, sign) where canonical_blade is
            ↳  the basis blade with indices
            sorted in ascending order and sign is either +1 or -1
            ↳  depending on the swap parity.

    This function uses a simple bubble sort algorithm to reorder the
    ↳  indices. Each time a swap
    occurs, the sign is flipped to reflect the antisymmetry of the
    ↳  basis blades.
    """
    canonical_blade = list(blade)
    sign = 1
    n = len(canonical_blade)
    for i in range(n):
        for j in range(0, n - i - 1):
            if canonical_blade[j] > canonical_blade[j + 1]:
                canonical_blade[j], canonical_blade[j + 1] =
                ↳  canonical_blade[j + 1], canonical_blade[j]
                sign = -sign
    return tuple(canonical_blade), sign

def geometric_product_blades(blade1, blade2, metric):
    """
    Compute the geometric product of two basis blades.

    This function concatenates the indices of the input blades,
    ↳  reorders them
    into canonical order (tracking the sign due to swaps), and then
    ↳  applies metric
    contractions when duplicate indices appear. In geometric
    ↳  algebra, repeated indices
    contract according to the metric values.

    Parameters:
        blade1 : tuple
            A tuple representing the first basis blade.
```

431

```
        blade2 : tuple
            A tuple representing the second basis blade.
        metric : dict
            A dictionary mapping basis indices to their metric
            ↪   values.

    Returns:
        tuple:
            (result_blade, factor) where:
              - result_blade is the resulting canonical basis blade
              ↪   after contracting duplicates.
              - factor is the overall multiplicative factor,
              ↪   including both the sign from reordering
                and the metric factors from contractions.
    """
    # Concatenate the two blades.
    combined = blade1 + blade2
    # Reorder the indices to canonical form.
    canonical, sign = symbolic_reorder_blade(combined)
    total_factor = sign
    canonical_list = list(canonical)
    i = 0
    # Process duplicate indices via metric contraction.
    while i < len(canonical_list) - 1:
        if canonical_list[i] == canonical_list[i + 1]:
            # When two equal indices are contracted, they contribute
            ↪   a factor from the metric.
            total_factor *= metric.get(canonical_list[i], 1)
            # Remove the contracted pair.
            del canonical_list[i:i + 2]
            # Do not increment i, as the list has shrunk.
        else:
            i += 1
    return tuple(canonical_list), total_factor

def geometric_product_multivectors(M1, M2, metric):
    """
    Compute the geometric product of two multivectors.

    Multivectors are represented as dictionaries mapping basis
    ↪   blades (tuples) to coefficients.
    The geometric product is computed by iterating over all pairs of
    ↪   terms from the two multivectors,
    multiplying their coefficients and basis blades, and then
    ↪   aggregating terms with identical blades.

    Parameters:
        M1 : dict
            The first multivector, with keys as blades (tuples) and
            ↪   values as coefficients.
        M2 : dict
            The second multivector in the same format.
        metric : dict
```

```
            The metric dictionary for computing products of basis
         ↪  vectors.

Returns:
    dict: The resulting multivector from the product M1*M2.
"""
result = {}
for blade1, coeff1 in M1.items():
    for blade2, coeff2 in M2.items():
        new_blade, factor = geometric_product_blades(blade1,
         ↪  blade2, metric)
        new_coeff = coeff1 * coeff2 * factor
        # Aggregate coefficients for like blades.
        if new_blade in result:
            result[new_blade] += new_coeff
        else:
            result[new_blade] = new_coeff
return result

def pretty_print_multivector(M):
    """
    Pretty prints a multivector in a readable canonical form.

    The multivector is displayed as a sum of terms where each term
     ↪  consists of a coefficient
    and its associated basis blade (e.g., e1e2 for a blade (1,2)). A
     ↪  scalar term is represented by '1'.

    Parameters:
        M : dict
            A multivector represented as a dictionary with keys as
             ↪  blade tuples and values as coefficients.
    """
    terms = []
    for blade, coeff in sorted(M.items()):
        if blade:
            blade_str = ''.join(f"e{idx}" for idx in blade)
        else:
            blade_str = "1"  # Scalar part if the blade is empty.
        terms.append(f"{coeff:+} {blade_str}")
    print(" ".join(terms))

# Example usage:
# Define two multivectors A and B.
# Here A = 3 + 2e1 + 1e2e3 and B = 1 + 1e1.
A = {
    (): 3,       # Scalar component: 3
    (1,): 2,     # Vector component: 2e1
    (2, 3): 1    # Bivector component: e2e3
}

B = {
    (): 1,    # Scalar component: 1
```

433

```
        (1,): 1    # Vector component: e1
}

# Compute the geometric product of A and B.
product = geometric_product_multivectors(A, B, metric)

# Print the result.
print("Geometric Product of A and B:")
pretty_print_multivector(product)

if __name__ == "__main__":
    # Run the example when executed as a standalone script.
    print("\nRunning example computation for the geometric product
    ↪  in Geometric Algebra:")
    print("Multivector A:")
    pretty_print_multivector(A)
    print("Multivector B:")
    pretty_print_multivector(B)
    product = geometric_product_multivectors(A, B, metric)
    print("Result of A * B:")
    pretty_print_multivector(product)
```

Chapter 62

Visualization of Multivectors: Techniques for Physicists

Foundations of Multivector Visualization

Multivectors, as composite entities built from scalars, vectors, bivectors, and higher-grade blades, present an abstract structure that often defies immediate intuitive interpretation. Their representation in an n-dimensional Euclidean space requires a careful synthesis of algebraic properties and geometric intuition. The formalism of geometric algebra expresses multivectors as linear combinations of basis blades, each of which carries orientation and magnitude information. A primary challenge is the conversion of the underlying algebraic complexity into visual constructs that preserve essential structural relationships. A robust visualization paradigm must encapsulate the inherent grading, the exterior product properties, and the metric signature associated with the geometric product.

Dimensional Reduction and Projection Techniques

Visualization of multivectors necessitates mapping high-dimensional algebraic constructs onto lower-dimensional graphical representations. In many cases, this process involves projecting individual blades onto a subspace that retains key relational features. Projection techniques, such as principal component analysis and orthogonal slicing, convert the multivector coordinates into a reduced coordinate system. The projection must account for the aggregation of geometric information where the coordinates are normalized and the influence of each blade is combined into a single set of spatial markers. Such an approach allows for the depiction of rotational invariance and other symmetries innate to the multivector structure.

For example, given a basis blade represented by a tuple of indices, each index can be mapped to a canonical axis in the target space. The resulting coordinates may be computed as the average of the contributions from each index, weighted according to the transformation rules of the multivector components. This process involves both algebraic manipulation and subsequent geometric interpretation.

Computational Techniques for Rendering Multivectors

The implementation of visualization methods involves algorithmic treatment of symbolic multivector expressions to yield numerical coordinates suitable for graphical rendering. Central to this procedure is the conversion of algebraic basis representations into spatial coordinates that reflect both magnitude and orientation. A function that performs this conversion must take into account the dimensionality of the ambient space and the composition of the multivector. The resulting coordinate system may be used to generate scatter plots, vector field diagrams, or more complex graphical constructs that are amenable to analysis via standard computer graphics frameworks.

An illustrative Python function, presented in a minted snippet, computes the coordinates associated with a given basis blade. This function maps the indices of the blade to their respective positions in a reduced-dimensional Euclidean space while normalizing the contributions. The function is written to handle blades of arbitrary grade, thereby serving as a fundamental building block for assembling comprehensive visual representations.

```
def compute_blade_coordinates(blade, dimension):
    """
    Compute coordinates for visualizing a basis blade in an
    ↪ n-dimensional Euclidean space.

    The function maps a basis blade, represented as a tuple of
    ↪ indices, to a set of coordinates
    in a reduced-dimensional space suitable for graphical rendering.
    ↪ Each index corresponds to an
    axis in the original space, and the function aggregates these
    ↪ contributions by normalizing their
    cumulative effect. For a multivector component with multiple
    ↪ indices, the resulting coordinate
    approximates the average direction inherent to the blade's
    ↪ geometric orientation.

    Parameters:
        blade : tuple
            A tuple of integers representing the indices of the
            ↪ basis blade.
        dimension : int
            The number of dimensions in the original Euclidean
            ↪ space.

    Returns:
        list:
            A list of floats representing the normalized coordinates
            ↪ in the reduced-dimensional space.
    """
    if not blade:
        return [0.0] * dimension
    coords = [0.0] * dimension
    for i in blade:
        if 1 <= i <= dimension:
            coords[i - 1] += 1.0
    norm = sum(coords)
    return [x / norm for x in coords] if norm != 0 else coords
```

Integration with Visualization Frameworks

The effective rendering of multivector visualizations further depends on the seamless integration of computational algorithms with graphical libraries. Libraries such as Matplotlib or more interactive frameworks like Plotly provide the necessary tools for plotting the coordinates computed from multivector components. In this integration, data structures that store the spatial representations of each blade are processed for visual output using scatter plots, vector arrows, or colored surfaces that denote different grades.

In addition to basic plotting, advanced techniques include the use of color mapping to represent additional scalar characteristics, interactive rotation to better understand the geometric orientation, and the overlay of multiple multivector components to reveal algebraic dependencies. These computational strategies enable a comprehensive visual analysis where abstract algebraic structures are translated into interpretable graphical forms, facilitating a deeper understanding of the interplay between geometry and algebra in physical models.

Python Code Snippet

```
import numpy as np
import matplotlib.pyplot as plt
from sklearn.decomposition import PCA

def compute_blade_coordinates(blade, dimension):
    """
    Compute coordinates for visualizing a basis blade in an
    ↪ n-dimensional Euclidean space.

    The function maps a basis blade, represented as a tuple of
    ↪ indices, to a set of coordinates
    in a reduced-dimensional space suitable for graphical rendering.
    ↪ Each index corresponds to an
    axis in the original space, and the function aggregates these
    ↪ contributions by normalizing their
    cumulative effect. For a multivector component with multiple
    ↪ indices, the resulting coordinate
    approximates the average direction inherent to the blade's
    ↪ geometric orientation.

    Parameters:
        blade : tuple
            A tuple of integers representing the indices of the
            ↪ basis blade.
        dimension : int
            The number of dimensions in the original Euclidean
            ↪ space.

    Returns:
        list:
```

```
            A list of floats representing the normalized coordinates
            ↪  in the reduced-dimensional space.
    """
    if not blade:
        return [0.0] * dimension
    coords = [0.0] * dimension
    for i in blade:
        if 1 <= i <= dimension:
            coords[i - 1] += 1.0
    norm = sum(coords)
    return [x / norm for x in coords] if norm != 0 else coords

def multivector_to_coordinates(multivector, dimension):
    """
    Convert a multivector represented as a dictionary of basis
    ↪  blades and their coefficients
    into a set of spatial coordinates for visualization.

    Parameters:
        multivector : dict
            A dictionary where keys are tuples representing basis
            ↪  blades and values are numerical coefficients.
        dimension : int
            The number of dimensions in the ambient Euclidean space.

    Returns:
        tuple:
            A tuple containing:
            - A NumPy array of coordinates for each blade.
            - A list of string labels representing each blade.
            - A list of corresponding weight coefficients.
    """
    coord_list = []
    labels = []
    weights = []
    for blade, weight in multivector.items():
        coords = compute_blade_coordinates(blade, dimension)
        # Scale the computed coordinates by the weight of the
        ↪  multivector component.
        scaled_coords = [c * weight for c in coords]
        coord_list.append(scaled_coords)
        labels.append(str(blade))
        weights.append(weight)
    return np.array(coord_list), labels, weights

def project_coordinates(coordinates, target_dim=2):
    """
    Reduce the dimensionality of the coordinate data to a target
    ↪  dimension using Principal Component Analysis (PCA).

    Parameters:
        coordinates : numpy.ndarray
```

```
            The original coordinate array with shape (num_blades,
            ↪ original_dimension).
        target_dim : int
            The target dimension for the projection (default is 2).

    Returns:
        numpy.ndarray:
            The coordinate array after dimensionality reduction,
            ↪ with shape (num_blades, target_dim).
    """
    if coordinates.shape[1] > target_dim:
        pca = PCA(n_components=target_dim)
        reduced_coords = pca.fit_transform(coordinates)
        return reduced_coords
    else:
        return coordinates

def plot_multivector(coordinates, labels):
    """
    Plot the multivector coordinates in a 2D scatter plot, labeling
    ↪ each point with its corresponding basis blade.

    Parameters:
        coordinates : numpy.ndarray
            A 2D array of coordinates for each multivector
            ↪ component.
        labels : list
            List of labels (as strings) corresponding to each
            ↪ coordinate.
    """
    plt.figure(figsize=(8, 6))
    plt.scatter(coordinates[:, 0], coordinates[:, 1], color='blue',
    ↪ s=100)
    for i, label in enumerate(labels):
        plt.annotate(label, (coordinates[i, 0], coordinates[i, 1]),
                     textcoords="offset points", xytext=(5, 5),
                     ↪ ha='center')
    plt.title("Multivector Visualization (PCA reduced to 2D)")
    plt.xlabel("Principal Component 1")
    plt.ylabel("Principal Component 2")
    plt.grid(True)
    plt.show()

def main():
    # Define the ambient space dimension.
    ambient_dimension = 4

    # Define a sample multivector with various blades and
    ↪ coefficients.
    multivector = {
        (): 1.0,            # Scalar component
        (1,): 2.0,          # Vector along axis 1
        (2,): -1.5,         # Vector along axis 2
```

```
        (3,): 0.8,          # Vector along axis 3
        (4,): -1.2,         # Vector along axis 4
        (1,2): 3.0,         # Bivector component in plane 1-2
        (2,3): -2.5,        # Bivector component in plane 2-3
        (3,4): 1.5          # Bivector component in plane 3-4
    }

    # Compute the coordinates for each basis blade in the
    ↪ multivector.
    coords, labels, weights =
    ↪ multivector_to_coordinates(multivector, ambient_dimension)

    # Project the coordinates to 2D for visualization.
    coords_2d = project_coordinates(coords, target_dim=2)

    # Render the multivector visualization using a 2D scatter plot.
    plot_multivector(coords_2d, labels)

if __name__ == "__main__":
    main()
```

Chapter 63

Multidimensional Rotations: Exploring Higher Dimensions

Mathematical Foundations of Higher-Dimensional Rotations

Rotations in spaces of dimension greater than three extend the classical notion of planar or spatial rotations into a framework where rotations occur within two-dimensional planes embedded in an n-dimensional Euclidean space. In the language of geometric algebra, a rotation is generated by a rotor, which is constructed as the exponential of a bivector. For a normalized bivector B, the rotor is expressed as

$$R = \exp\left(-\frac{B\theta}{2}\right),$$

where θ is the rotation angle. The action of a rotor on a vector v is implemented via the sandwiching operation

$$v' = R v R^\dagger,$$

with $R^\dagger$ denoting the reverse of R. This formulation consolidates the description of rotations across arbitrarily many planes, thereby permitting a unified treatment of multidimensional rotational dynamics under the framework of geometric algebra.

Representation of Higher-Dimensional Rotors in Geometric Algebra

The representation of rotors in higher dimensions capitalizes on the inherent properties of bivectors to define oriented planes. Unlike conventional rotation matrices, which require complex parameterizations involving several angles (as in the case of Euler angles in three dimensions), rotors offer a compact and elegant alternative. In this setting, the bivector encapsulates the information regarding the plane of rotation, and the associated rotor encodes both the rotational magnitude and the orientation of that plane. By parameterizing rotations as an exponential of a bivector, it becomes possible to efficiently compute the transformation of vectors and multivectors in a high-dimensional setting. The underlying computation reduces to the evaluation of trigonometric functions and the normalization of the bivector, thereby easing the numerical implementation of such rotations.

Algorithmic Construction of Multidimensional Rotations

The numerical construction of multidimensional rotations relies on the explicit computation of the rotor from a given bivector and a prescribed rotation angle. The rotor can be constructed by first normalizing the bivector and then evaluating the cosine and sine components at half the rotation angle. The resulting rotor adopts the form

$$R = \cos\left(\frac{\theta}{2}\right) - B_{\text{norm}} \sin\left(\frac{\theta}{2}\right),$$

where B_{norm} is the normalized bivector. The following Python function implements this computation. The function accepts a numerical representation of the bivector (typically as a NumPy array) along with the rotation angle in radians, and returns the scalar and bivector parts of the rotor necessary for constructing the sandwich product.

```
def compute_rotor(bivector, theta):
    """
    Compute a rotor for a rotation in a plane defined by the
    ↪ bivector in higher dimensions.
```

```
    This function calculates the rotor R corresponding to a rotation
 ↪  by an angle theta.
    The rotor is based on the exponential mapping R = exp(-B * theta
 ↪  / 2), where B is required
    to be normalized. It returns a tuple consisting of the scalar
 ↪  component, given by cos(theta/2),
    and the bivector component, which is the normalized bivector
 ↪  multiplied by sin(theta/2).

    Parameters:
        bivector : numpy.ndarray
            A NumPy array representing the bivector that defines the
         ↪  plane of rotation.
        theta : float
            The rotation angle in radians.

    Returns:
        tuple:
            A tuple (scalar, bivector_part) where 'scalar'
         ↪  represents cos(theta/2) and 'bivector_part'
            represents sin(theta/2) * (normalized bivector),
         ↪  encapsulating the necessary information
            for performing the rotor sandwich product.
    """
    normB = np.linalg.norm(bivector)
    if normB == 0:
        return (1.0, np.zeros_like(bivector))   # Identity rotor for
     ↪  zero bivector
    normalized_B = bivector / normB
    return (np.cos(theta/2), np.sin(theta/2) * normalized_B)
```

This algorithmic approach leverages the structure of geometric algebra to reduce the computational overhead typically associated with multidimensional rotations. By simplifying the rotation to the evaluation of sine and cosine functions at half the angle and by normalizing the bivector, the method naturally accommodates the complexity introduced by higher dimensions while maintaining numerical efficiency.

Python Code Snippet

```
import numpy as np

def compute_rotor(bivector, theta):
    """
    Compute a rotor for a rotation in a plane defined by the
 ↪  bivector in higher dimensions.
```

```
    This function calculates the rotor R corresponding to a rotation
    ↪ by an angle theta.
    The rotor is based on the exponential mapping:
        R = exp(-B * theta/2) = cos(theta/2) - B_norm * sin(theta/2)
    for a normalized bivector B_norm.

    Parameters:
        bivector : numpy.ndarray
            A NumPy array representing the bivector (via its
            ↪ components) that defines the plane of rotation.
        theta : float
            The rotation angle in radians.

    Returns:
        tuple:
            A tuple (scalar, bivector_part) representing the rotor.
            Here, scalar = cos(theta/2) and bivector_part = -
            ↪ sin(theta/2) * (normalized bivector).
    """
    normB = np.linalg.norm(bivector)
    if normB == 0:
        return (1.0, np.zeros_like(bivector))  # Identity rotor for
        ↪ zero bivector
    normalized_B = bivector / normB
    return (np.cos(theta/2), -np.sin(theta/2) * normalized_B)

def rotor_reverse(rotor):
    """
    Compute the reverse (reversion) of a rotor.

    For a rotor R represented as (a, B), where a is the scalar part
    ↪ and B is the bivector part,
    its reverse R† is given by (a, -B).

    Parameters:
        rotor : tuple
            A tuple (scalar, bivector_part) representing the rotor.

    Returns:
        tuple:
            The reverse of the rotor, (scalar, -bivector_part).
    """
    a, B = rotor
    return (a, -B)

def rotate_vector_in_plane(v, e1, e2, theta):
    """
    Rotate a vector v in the plane defined by the orthonormal basis
    ↪ vectors e1 and e2.

    This function simulates the rotor sandwich product R v R† by
    ↪ decomposing v into the
```

445

```
    rotation plane and its orthogonal complement. The component in
 ↪   the plane is rotated by
    the angle theta, while the perpendicular component remains
 ↪   unchanged.

    Parameters:
        v : numpy.ndarray
            The vector to be rotated.
        e1, e2 : numpy.ndarray
            Orthonormal vectors that define the rotation plane.
        theta : float
            The rotation angle in radians.

    Returns:
        numpy.ndarray:
            The rotated vector.
    """
    # Project v onto the rotation plane spanned by e1 and e2.
    v_parallel = np.dot(v, e1) * e1 + np.dot(v, e2) * e2
    # Extract the component orthogonal to the rotation plane.
    v_perp = v - v_parallel
    # Compute coefficients along the plane basis directions.
    a = np.dot(v, e1)
    b = np.dot(v, e2)
    # Rotate the parallel component using the standard 2D rotation
 ↪   formula.
    rotated_parallel = (a * np.cos(theta) - b * np.sin(theta)) * e1
 ↪   \
                     + (a * np.sin(theta) + b * np.cos(theta)) * e2
    return v_perp + rotated_parallel

# Example usage demonstrating the rotor-based rotation in a
 ↪ higher-dimensional context.
if __name__ == "__main__":
    # Define two orthonormal vectors e1 and e2 that span the
 ↪   rotation plane in R^n (here, n = 3 for demonstration).
    e1 = np.array([1.0, 0.0, 0.0])
    e2 = np.array([0.0, 1.0, 0.0])

    # For a bivector representing the plane spanned by e1 and e2,
 ↪   one common approach in 3D is to use
    # the duality between bivectors and vectors. Here, we adopt the
 ↪   representation of the bivector as:
    bivector = np.array([0.0, 0.0, 1.0])
    # Note: In higher dimensions, bivectors are more naturally
 ↪   represented as antisymmetric arrays or
    # using basis blades. This example uses a 3D simplification.

    # Define the rotation angle (e.g., 45 degrees).
    theta = np.pi / 4   # 45 degrees in radians

    # Compute the rotor from the given bivector and angle.
    rotor = compute_rotor(bivector, theta)
```

```python
print("Rotor (scalar, bivector_part):", rotor)

# Compute the reverse of the rotor.
rotor_rev = rotor_reverse(rotor)
print("Rotor Reverse:", rotor_rev)

# Define a sample vector to be rotated.
v = np.array([1.0, 1.0, 0.0])
print("Original Vector:", v)

# Rotate the vector using the GA-inspired rotor sandwich
↪   approach.
# This function rotates only the component of the vector within
↪   the plane spanned by e1 and e2.
v_rotated = rotate_vector_in_plane(v, e1, e2, theta)
print("Rotated Vector:", v_rotated)
```

Chapter 64

Tensor and Multivector Interplay: Algebraic Connections

Mathematical Foundations

Tensors establish a rigorous framework for encoding multilinear relationships in physical systems. In contrast, multivectors, the fundamental constructs of geometric algebra, encompass quantities ranging from scalars to pseudoscalars in an integrated algebraic structure. A rank-2 tensor T_{ij}, when decomposed into its symmetric and antisymmetric components, reveals an intrinsic connection to multivectors. In particular, the antisymmetric component

$$\frac{1}{2}(T_{ij} - T_{ji})$$

exhibits a natural correspondence with a bivector. Representing oriented plane segments in an n-dimensional Euclidean space, a bivector B is expressed in a basis $\{e_i\}$ as

$$B = \frac{1}{2}\sum_{i,j} B_{ij}\, e_i \wedge e_j,$$

with the wedge product ensuring that $B_{ij} = -B_{ji}$. Such representations form the foundation for applications in electromagnetism, relativity, and other advanced fields in physics.

Algebraic Mappings and Interplay

The mapping between tensor and multivector formulations is established by identifying components of antisymmetric tensors with corresponding graded elements in geometric algebra. An antisymmetric tensor A_{ij} finds its counterpart in the bivector

$$B = \frac{1}{2} \sum_{i,j} A_{ij}\, e_i \wedge e_j,$$

which preserves the geometric orientation and magnitude encoded in the tensor components. This equivalence not only streamlines the representation of physical quantities but also simplifies the formulation of transformation rules, such as rotations and boosts, as the operations in geometric algebra are inherently coordinate-free.

In higher-dimensional settings, the interconversion of complex tensorial relationships into multivector expressions lends itself to more concise analytical expressions and efficient computational algorithms. The algebraic structure of multivectors consolidates tensor operations—such as index contraction and outer products—into a unified framework that readily accommodates the transformation properties under changes of basis.

Computational Conversion and Implementation

Efficient numerical strategies often require converting an antisymmetric tensor, stored as a two-dimensional array, into its corresponding bivector form. The conversion process leverages the inherent antisymmetry to gather only the unique components. The function below demonstrates a procedure that maps a rank-2 antisymmetric tensor into a one-dimensional array representing the independent bivector components. Consistency in the ordering is essential and must align with the chosen multivector basis for subsequent algebraic manipulations.

```
def tensor_to_bivector(antisym_tensor):
    """
    Convert an antisymmetric tensor represented as a 2D NumPy array
    into a bivector component vector. In an n-dimensional space, the
    tensor is antisymmetric, i.e., A[i, j] = -A[j, i]. The bivector
```

```
is represented as a 1D array containing the independent
    components.

Parameters:
    antisym_tensor (numpy.ndarray): A square 2D array of shape
        (n, n)
        representing the antisymmetric tensor.

Returns:
    numpy.ndarray: A 1D array with length n*(n-1)/2 holding the
        unique
        bivector components corresponding to the tensor.
"""
import numpy as np
n = antisym_tensor.shape[0]
bivector_components = []
for i in range(n):
    for j in range(i + 1, n):
        bivector_components.append(antisym_tensor[i, j])
return np.array(bivector_components)
```

The algorithm systematically iterates over the index pairs (i, j) with $i < j$, thereby eliminating redundant contributions. The resulting array can be directly employed in multivector computations where operations such as the geometric product or the evaluation of exponential maps are executed.

Interconversion of Representational Formalisms

In analytical and numerical studies, the interconversion between tensorial and multivector representations enhances both clarity and performance. Contraction of indices in tensor algebra translates to inner products within the geometric algebra framework, while the exterior (wedge) product perfectly encapsulates the antisymmetric properties inherent in tensors. This correspondence ensures that the invariance under coordinate transformations is automatically preserved when switching between representations.

The algebraic advantages of multivectors become particularly evident in scenarios where complex physical phenomena, such as stress distributions or curvature tensors, are involved. The compact encoding of information permits the development of algorithms that are less susceptible to the numerical instabilities often encountered in high-dimensional tensor operations. Consequently, the interplay between tensors and multivectors forms a powerful tool in the modeling and simulation of advanced physical systems.

Python Code Snippet

```
import numpy as np

def tensor_to_bivector(antisym_tensor):
    """
    Convert an antisymmetric tensor represented as a 2D NumPy array
    into a bivector component vector. In an n-dimensional space, the
    tensor is antisymmetric, i.e., A[i, j] = -A[j, i]. The bivector
    is represented as a 1D array containing the unique components.

    According to the formula:
        B = 1/2 * sum_{i,j} A_{ij} e_i e_j
    we only consider the unique components where i < j.

    Parameters:
        antisym_tensor (numpy.ndarray): A square 2D array of shape
        ↪ (n, n)
            representing the antisymmetric tensor.

    Returns:
        numpy.ndarray: A 1D array with length n*(n-1)/2 holding the
        ↪ unique
            bivector components corresponding to the tensor.
    """
    n = antisym_tensor.shape[0]
    bivector_components = []
    for i in range(n):
        for j in range(i + 1, n):
            bivector_components.append(antisym_tensor[i, j])
    return np.array(bivector_components)

def bivector_to_tensor(bivector, n):
    """
    Reconstruct the full antisymmetric tensor from its bivector
    ↪ component vector.

    Given the ordering of unique bivector components as:
        [A[0,1], A[0,2], ..., A[0,n-1], A[1,2], ..., A[n-2, n-1]]
    we rebuild the tensor as:
        T[i, j] = value for i < j and T[j, i] = -value.

    Parameters:
        bivector (numpy.ndarray): A 1D array with length n*(n-1)/2
        ↪ containing
            the unique bivector components.
```

```
            n (int): Dimension of the original space (resulting in an n
        ↪    x n tensor).

    Returns:
        numpy.ndarray: The reconstructed antisymmetric tensor of
        ↪    shape (n, n).
    """
    antisym_tensor = np.zeros((n, n))
    k = 0
    for i in range(n):
        for j in range(i + 1, n):
            antisym_tensor[i, j] = bivector[k]
            antisym_tensor[j, i] = -bivector[k]
            k += 1
    return antisym_tensor

def construct_random_antisym_tensor(n):
    """
    Construct a random antisymmetric tensor of dimension n x n.

    This function generates an n x n matrix with random entries and
    ↪    enforces
    the antisymmetry property by computing A - A^T.

    Parameters:
        n (int): Dimension of the tensor.

    Returns:
        numpy.ndarray: A random antisymmetric tensor of shape (n,
        ↪    n).
    """
    A = np.random.rand(n, n)
    antisym_tensor = A - A.T
    return antisym_tensor

def compute_bivector_norm(bivector):
    """
    Compute the norm of a bivector component vector.

    This computes the Euclidean norm (L2 norm) which can serve as a
    ↪    measure
    of the magnitude of the bivector, reflecting the overall
    ↪    strength of the
    antisymmetric component.

    Parameters:
        bivector (numpy.ndarray): A 1D array of bivector components.

    Returns:
        float: The Euclidean norm of the bivector.
    """
    return np.linalg.norm(bivector)
```

452

```python
def main():
    """
    Main routine to demonstrate the conversion of an antisymmetric
    ↪   tensor
    to bivector components and its reconstruction. This example
    ↪   outlines the important
    computation steps discussed in the chapter, including:
       - Constructing an antisymmetric tensor.
       - Converting the tensor to its unique bivector representation.
       - Computing the norm of the bivector.
       - Reconstructing the tensor from the bivector components.
       - Verifying the reconstruction accuracy.
    """
    # Set the dimension of the space (example: 4-dimensional)
    n = 4

    # Generate a random antisymmetric tensor
    tensor = construct_random_antisym_tensor(n)
    print("Original Antisymmetric Tensor (n = {}):".format(n))
    print(tensor)

    # Convert the tensor to bivector components
    bivector = tensor_to_bivector(tensor)
    print("\nBivector Components:")
    print(bivector)

    # Compute the norm of the bivector for further geometric
    ↪   analysis
    norm = compute_bivector_norm(bivector)
    print("\nBivector Norm (L2 norm): {:.4f}".format(norm))

    # Reconstruct the tensor from the bivector components
    reconstructed_tensor = bivector_to_tensor(bivector, n)
    print("\nReconstructed Tensor from Bivector Components:")
    print(reconstructed_tensor)

    # Calculate and display the reconstruction error (should be near
    ↪   zero)
    error = np.linalg.norm(tensor - reconstructed_tensor)
    print("\nReconstruction Error (Frobenius norm):
    ↪   {:.2e}".format(error))

if __name__ == "__main__":
    main()
```

Chapter 65

Data Structures in GA: Efficient Computational Representations

Design Principles and Memory Layout

In the realm of geometric algebra (GA), the design of data structures involves a careful balance between mathematical expressiveness and computational efficiency. A multivector, which may contain a scalar, vector, bivector, and higher-grade elements, is naturally represented as an associative collection of coefficients indexed by basis blades. The basis blades are often encoded as ordered tuples or bit masks, thereby preserving the intrinsic ordering and antisymmetry inherent in GA. In implementations targeting high-performance applications, data organization is prioritized to facilitate rapid access, reduce memory overhead, and optimize arithmetic operations such as the geometric product. For a space of dimension n, a fully stored multivector exhibits a size proportional to 2^n, which necessitates a sparse or hierarchical representation in many computational scenarios.

Mapping Algebraic Structures to Data Types

The storage of GA elements in software typically leverages standard data types. Adopting built-in structures such as Python dictionaries enables the mapping from a basis blade, represented by an immutable tuple of indices, to its corresponding scalar coefficient. Such a representation is particularly useful in the context of sparse multivectors where many coefficients are identically zero. Furthermore, dense representations using ordered arrays may be advantageous for low-dimensional spaces or when vectorized operations are applicable. The design must account for both the algebraic operations defined on multivectors and the native memory layout of the computing environment in order to exploit cache locality and parallel processing capabilities.

Implementation of Custom GA Data Structures

A central operation in GA computations involves the initialization of multivector objects from a set of basis components. The following Python function illustrates a method for constructing a multivector using a dictionary, where the keys represent basis blades and the corresponding values are the scalar coefficients. This approach encapsulates the mapping between the algebraic formalism and the computational representation, ensuring that the underlying data structure can be efficiently manipulated in subsequent arithmetic operations.

```
def create_multivector(components):
    """
    Create a geometric algebra multivector with specified
    ↪ components.

    The multivector is represented as a dictionary where each key is
    ↪ a tuple
    corresponding to a basis blade (e.g., (1,), (1,2), (2,3)) and
    ↪ each value is the
    numerical coefficient associated with that blade. This
    ↪ representation affords
    a sparse storage mechanism particularly suited for
    ↪ high-dimensional GA spaces.

    Parameters:
```

```
            components (dict): A mapping from tuples representing basis
        ↪    blades to
                         numeric coefficients.

    Returns:
        dict: A multivector represented as an associative array.
    """
    multivector = {}
    for blade, coefficient in components.items():
        multivector[blade] = coefficient
    return multivector
```

Efficiency and Optimization Considerations

The efficiency of GA operations hinges on the choice of data structures as much as on the algorithms that act upon them. When multivectors are stored as key-value pairs, the ordering of keys can be exploited to perform rapid look-ups and to reduce computational complexity during operations such as addition, contraction, and multiplication. An ordered layout may further facilitate the precomputation of multiplication tables or the application of partitioned arithmetic algorithms. In contrast, a dense, array-based representation provides benefits in terms of contiguous memory allocation and compatibility with vectorized numerical libraries, particularly for operations that are executed repeatedly or in a parallelized context.

Memory usage and runtime performance are profoundly influenced by the trade-offs between sparse and dense representations. In scenarios with a high prevalence of zero coefficients or where the GA elements mix grades in a non-uniform manner, a sparse dictionary-based implementation reduces unnecessary overhead. Conversely, in environments where every component is non-trivial and the dimension is modest, a dense array implementation may yield superior performance. Optimizing these representations often demands profiling and iterative refinement, taking into account processor cache behavior, the cost of memory lookups, and the efficiency of arithmetic operations under the specific constraints of the hardware.

Each data structure design must also be amenable to operations such as the geometric product, which require systematic iteration over pairs of basis blades. In this context, maintaining a coherent ordering of blades and minimizing the overhead of key comparisons are of paramount importance. The integrative design of the data structure, together with algorithmic improvements in multi-vector arithmetic, contributes significantly to robust and efficient implementations of geometric algebra in computational physics and computer science.

Python Code Snippet

```python
from collections import defaultdict
import itertools

def grade(blade):
    """
    Return the grade (i.e., the number of basis vectors) of the
    ↪ blade.
    """
    return len(blade)

def canonical_blade(blade):
    """
    Returns the canonical form of a basis blade along with the sign
    ↪ resulting
    from reordering. The blade is given as a tuple of indices.

    Algorithm:
        1. Convert the blade to a list.
        2. Sort the list while counting the number of swaps (which
    ↪ gives the sign).
        3. Remove adjacent duplicate indices (recalling that e_i*e_i =
    ↪ 1 in Euclidean GA).

    Parameters:
        blade (tuple): A tuple representing a product of basis
    ↪ vectors.

    Returns:
        tuple: (canonical_blade, sign) where canonical_blade is a
    ↪ sorted tuple
               with canceling pairs removed, and sign is +1 or -1.
    """
    blade_list = list(blade)
    sign = 1
    n = len(blade_list)
    # Bubble sort to count the number of transpositions needed.
    for i in range(n):
        for j in range(i + 1, n):
            if blade_list[i] > blade_list[j]:
```

```python
                    blade_list[i], blade_list[j] = blade_list[j],
                    ↪   blade_list[i]
                    sign *= -1
    # Remove pairs of identical indices (e_i * e_i = 1)
    result = []
    i = 0
    while i < len(blade_list):
        if i + 1 < len(blade_list) and blade_list[i] == blade_list[i
        ↪   + 1]:
            # The pair cancels out
            i += 2
        else:
            result.append(blade_list[i])
            i += 1
    return tuple(result), sign

def geometric_product_of_blades(blade1, blade2):
    """
    Compute the geometric product of two basis blades.

    The geometric product of basis blades can be computed by:
      1. Concatenating the two blades.
      2. Converting the result into its canonical representation.

    Parameters:
        blade1 (tuple): A tuple representing the first basis blade.
        blade2 (tuple): A tuple representing the second basis blade.

    Returns:
        tuple: (result_blade, sign) where result_blade is the
        ↪   canonical basis blade tuple,
              and sign is the cumulative sign change from
              ↪   reordering.
    """
    combined = blade1 + blade2
    canonical, sign = canonical_blade(combined)
    return canonical, sign

class Multivector:
    """
    Class representing a multivector in geometric algebra using a
    ↪   sparse dictionary-based
    storage. Each key in the dictionary is a tuple representing a
    ↪   basis blade (e.g., (), (1,),
    (1,2), etc.), and the corresponding value is its numerical
    ↪   coefficient.
    """
    def __init__(self, components=None):
        """
        Initialize the multivector. If components is provided, it
        ↪   should be a dictionary
        mapping tuples (basis blades) to numerical coefficients.
        """
```

```python
        if components is None:
            self.components = {}
        else:
            # Filter out components with zero coefficient for
            ↪ sparsity.
            self.components = {blade: coeff for blade, coeff in
            ↪ components.items() if coeff != 0}

    @staticmethod
    def from_components(components):
        """
        Static method to create a Multivector from a dictionary of
        ↪ components.

        Parameters:
            components (dict): Mapping from basis blade tuples to
            ↪ coefficients.

        Returns:
            Multivector: An instance representing the given
            ↪ multivector.
        """
        return Multivector(components)

    def __add__(self, other):
        """
        Add two multivectors.
        """
        result = defaultdict(float)
        for blade, coeff in self.components.items():
            result[blade] += coeff
        for blade, coeff in other.components.items():
            result[blade] += coeff
        # Filter out near-zero components.
        return Multivector({blade: coeff for blade, coeff in
        ↪ result.items() if abs(coeff) > 1e-12})

    def __sub__(self, other):
        """
        Subtract two multivectors.
        """
        result = defaultdict(float)
        for blade, coeff in self.components.items():
            result[blade] += coeff
        for blade, coeff in other.components.items():
            result[blade] -= coeff
        return Multivector({blade: coeff for blade, coeff in
        ↪ result.items() if abs(coeff) > 1e-12})

    def __mul__(self, other):
        """
        Overload the multiplication operator.
        Supports:
```

```
            - Scalar multiplication if 'other' is an int or float.
            - Geometric product if 'other' is another Multivector.
        """
        if isinstance(other, (int, float)):
            return Multivector({blade: coeff * other for blade,
            ↪  coeff in self.components.items()})
        elif isinstance(other, Multivector):
            result = defaultdict(float)
            # Iterate over all pairs of basis blades from self and
            ↪  other.
            for blade1, coeff1 in self.components.items():
                for blade2, coeff2 in other.components.items():
                    new_blade, sign =
                    ↪  geometric_product_of_blades(blade1, blade2)
                    result[new_blade] += coeff1 * coeff2 * sign
            return Multivector({blade: coeff for blade, coeff in
            ↪  result.items() if abs(coeff) > 1e-12})
        else:
            raise TypeError("Multiplication with type {} not
            ↪  supported".format(type(other)))

    def __rmul__(self, other):
        """
        Enable scalar multiplication with the scalar on the left.
        """
        return self.__mul__(other)

    def __str__(self):
        """
        Return a human-readable string representation of the
        ↪  multivector.
        """
        if not self.components:
            return "0"
        terms = []
        # Sort blades first by grade then lexicographically.
        for blade in sorted(self.components.keys(), key=lambda x:
        ↪  (len(x), x)):
            coeff = self.components[blade]
            if blade == ():
                terms.append(f"{coeff:.3g}")
            else:
                basis = "".join(f"e{i}" for i in blade)
                terms.append(f"{coeff:.3g}{basis}")
        return " + ".join(terms)

    def __repr__(self):
        return f"Multivector({self.components})"

if __name__ == "__main__":
    # Example: Define multivectors in a 3-dimensional GA framework.

    # Scalar multivector (grade-0)
```

```
mv_scalar = Multivector.from_components({(): 3.0})

# Vector multivector (grade-1)
mv_vector = Multivector.from_components({(1,): 1.0, (2,): 2.0,
↪ (3,): 3.0})

# Bivector multivector (grade-2)
mv_bivector = Multivector.from_components({(1,2): 4.0, (2,3):
↪ 5.0})

print("Scalar Multivector:")
print(mv_scalar)

print("\nVector Multivector:")
print(mv_vector)

print("\nBivector Multivector:")
print(mv_bivector)

# Demonstrate addition: vector + bivector
mv_sum = mv_vector + mv_bivector
print("\nAddition (Vector + Bivector):")
print(mv_sum)

# Demonstrate geometric product: vector * bivector
mv_product = mv_vector * mv_bivector
print("\nGeometric Product (Vector * Bivector):")
print(mv_product)

# Demonstrate scalar multiplication
mv_scaled = 2 * mv_vector
print("\nScalar Multiplication (2 * Vector):")
print(mv_scaled)

# Chaining operations: (scalar + vector) * bivector
mv_chain = (mv_scalar + mv_vector) * mv_bivector
print("\nChained Operation ((Scalar + Vector) * Bivector):")
print(mv_chain)
```

Chapter 66

Simulation Techniques with GA: Tools for Physical Modeling

Mathematical Framework for GA-Based Simulations

Geometric algebra (GA) provides a unified language for describing physical systems, where physical quantities are expressed as multivectors capable of representing scalars, vectors, bivectors, and higher-grade elements. The operations defined in GA, such as the geometric product

$$uv = u \cdot v + u \wedge v,$$

preserve both metric and oriented geometric information. In the simulation of complex physical phenomena, a state is often encapsulated as a multivector $X(t)$ whose time evolution satisfies differential equations of the form

$$\frac{dX}{dt} = F(X(t), t),$$

where the function F involves GA operations and encodes the dynamics of the system. The inherent coordinate-free formulation and the preservation of geometric invariants make GA a robust tool for simulating systems with rotational symmetry, relativistic dynamics, or coupled geometric constraints.

Discretization and Time Integration in GA Simulations

The transition from continuous models to computable approximations requires a discretization of the governing differential equations. In GA-based simulations, discrete time-stepping schemes are applied to evolve the multivector state. A basic integration technique employs the explicit Euler method:

$$X(t + \Delta t) = X(t) + \Delta t \cdot F(X(t), t),$$

which serves as a prototype for more advanced methods such as Runge–Kutta or symplectic integrators that can be adapted to retain the algebraic structure of GA operations. The discretization process must account for the non-commutative nature of the geometric product and the graded structure of the multivectors. Efficient simulation requires that the data structures and numerical methods preserve the sparsity and hierarchical ordering inherent in GA representations, thereby ensuring both accuracy and computational efficiency.

Algorithmic Implementation Example

A central component in numerical simulations is a function that performs a single time-step update of the system's state. The following Python function outlines a basic Euler integration step for a system represented in GA. In this context, the current state is assumed to be encoded as a multivector, and the derivative function, defined through GA operations, computes the time derivative. The function returns the updated state calculated over an interval Δt.

```
def simulate_ga_time_step(state, t, dt, derivative_func):
    """
    Compute a single time step update for a system represented in
    ↪ geometric algebra.

    The function accepts the current state 'state' (a multivector),
    ↪ the current
    time 't', a time increment 'dt', and a function
    ↪ 'derivative_func' that computes
    the time derivative of the state based on GA operations. An
    ↪ explicit Euler method
    is used here, where the updated state is determined by:
```

```
        state_new = state + dt * derivative_func(state, t)

    Parameters:
        state: Multivector representing the current system state.
        t: Current time.
        dt: Time step increment.
        derivative_func: Function that returns the derivative of the
        ↪ state.

    Returns:
        The updated multivector state after the time step dt.
    """
    deriv = derivative_func(state, t)
    new_state = state + dt * deriv
    return new_state
```

The function above illustrates an integral component within a simulation routine. By encapsulating the time-stepping logic in a dedicated function and leveraging the algebraic operations intrinsic to GA, the simulation framework can maintain high fidelity to the continuous dynamics while benefiting from computationally efficient data structures. The use of GA allows the integration scheme to capture subtle geometric features of the physical system, such as conservation laws and symmetry properties, that may be obscured or lost in traditional vector calculus formulations.

Python Code Snippet

```
import math

class Multivector:
    """
    A simple implementation of a 2D geometric algebra (GA)
    ↪ multivector.
    In 2D GA, a multivector is represented as:
        M = a + b*e1 + c*e2 + d*e12
    where e1^2 = e2^2 = 1, e1e2 = -e2e1, and e12 = e1e2 with e12^2 =
    ↪ -1.
    """
    def __init__(self, a=0.0, b=0.0, c=0.0, d=0.0):
        self.a = a  # Scalar component
        self.b = b  # Coefficient for vector component e1
        self.c = c  # Coefficient for vector component e2
        self.d = d  # Coefficient for bivector component e12

    def __add__(self, other):
        if isinstance(other, Multivector):
```

```python
            return Multivector(self.a + other.a,
                               self.b + other.b,
                               self.c + other.c,
                               self.d + other.d)
        else:
            # Assume other is a scalar
            return Multivector(self.a + other, self.b, self.c,
            ↪    self.d)

    def __radd__(self, other):
        return self.__add__(other)

    def __sub__(self, other):
        if isinstance(other, Multivector):
            return Multivector(self.a - other.a,
                               self.b - other.b,
                               self.c - other.c,
                               self.d - other.d)
        else:
            return Multivector(self.a - other, self.b, self.c,
            ↪    self.d)

    def __rsub__(self, other):
        if isinstance(other, Multivector):
            return other.__sub__(self)
        else:
            return Multivector(other - self.a, -self.b, -self.c,
            ↪    -self.d)

    def __mul__(self, other):
        # Scalar multiplication
        if isinstance(other, (int, float)):
            return Multivector(self.a * other,
                               self.b * other,
                               self.c * other,
                               self.d * other)
        elif isinstance(other, Multivector):
            # Geometric product (GA product) for 2D
            # Let self = a + b*e1 + c*e2 + d*e12 and
            #     other = e + f*e1 + g*e2 + h*e12.
            # Using the multiplication rules:
            #   e1^2 = e2^2 = 1,  e1e2 = e12,  e2e1 = -e12,  e12^2 =
            ↪    -1,
            # we obtain:
            # Scalar      : a*e + b*f + c*g - d*h
            # e1 component : a*f + b*e + d*g - c*h
            # e2 component : a*g + c*e + b*h - d*f
            # e12 component: a*h + b*g - c*f + d*e
            a, b, c, d = self.a, self.b, self.c, self.d
            e, f, g, h = other.a, other.b, other.c, other.d
            scalar = a*e + b*f + c*g - d*h
            e1 = a*f + b*e + d*g - c*h
            e2 = a*g + c*e + b*h - d*f
```

```python
            e12 = a*h + b*g - c*f + d*e
            return Multivector(scalar, e1, e2, e12)
        else:
            raise TypeError("Multiplication with type {} not
            ↪    supported.".format(type(other)))

    def __rmul__(self, other):
        # For left-hand scalar multiplication.
        return self.__mul__(other)

    def __str__(self):
        parts = []
        if abs(self.a) > 1e-8:
            parts.append(f"{self.a:.2f}")
        if abs(self.b) > 1e-8:
            parts.append(f"{self.b:.2f}e1")
        if abs(self.c) > 1e-8:
            parts.append(f"{self.c:.2f}e2")
        if abs(self.d) > 1e-8:
            parts.append(f"{self.d:.2f}e12")
        if not parts:
            return "0"
        return " + ".join(parts)

def simulate_ga_time_step(state, t, dt, derivative_func):
    """
    Compute a single time step update for a system represented in
    ↪    geometric algebra.

    The integration is performed via the explicit Euler method:
        state_new = state + dt * derivative_func(state, t)

    Parameters:
        state           : Multivector representing the current system
        ↪    state.
        t               : Current time.
        dt              : Time step increment.
        derivative_func: Function that computes the derivative of
        ↪    the state.

    Returns:
        The updated Multivector state after a time step dt.
    """
    deriv = derivative_func(state, t)
    new_state = state + dt * deriv
    return new_state

def derivative_func(state, t):
    """
    Define the time derivative of the state in the GA simulation.

    For demonstration purposes, we simulate a rotational dynamic.
```

```
    In 2D GA, a rotation can be generated by a bivector. Here, we
    ↪ define:
        dX/dt = R * X
    where the constant rotor generator R = e12 (represented as
    ↪ Multivector(0,0,0,1))
    generates a rotation (90° per unit time in this simple example).
    """
    # Rotor generator R = e12
    R = Multivector(0.0, 0.0, 0.0, 1.0)
    return R * state

def run_simulation(initial_state, t0, dt, num_steps):
    """
    Run the GA-based simulation using the explicit Euler integration
    ↪ scheme.

    Parameters:
        initial_state: The initial state as a Multivector.
        t0           : Initial time.
        dt           : Time step size.
        num_steps    : Number of integration steps.

    Returns:
        A list of (time, state) tuples representing the system
        ↪ evolution.
    """
    states = []
    state = initial_state
    t = t0
    states.append((t, state))
    for _ in range(num_steps):
        state = simulate_ga_time_step(state, t, dt, derivative_func)
        t += dt
        states.append((t, state))
    return states

if __name__ == "__main__":
    # Example simulation: Rotate the unit vector along e1
    ↪ continuously.
    # Represent e1 as the multivector: 0 + 1*e1 + 0*e2 + 0*e12.
    initial_state = Multivector(0.0, 1.0, 0.0, 0.0)

    # Set simulation parameters.
    t0 = 0.0          # Initial time.
    dt = 0.1          # Time step size.
    num_steps = 20    # Total number of time steps.

    simulation_data = run_simulation(initial_state, t0, dt,
    ↪ num_steps)

    # Print the simulation results.
    print("Time\tState")
    for time_val, state in simulation_data:
```

```python
print(f"{time_val:.2f}\t{state}")
```